Theory of Electric and Magnetic Orientation in Sharks and Rays Revisited

Adrianus J. Kalmijn

Theory of Electric and Magnetic Orientation in Sharks and Rays Revisited

Physical Principles, Biological Evidence, and Rebuttal of Misconceptions

Springer

Adrianus J. Kalmijn
Scripps Institution of Oceanography, University of California
San Diego, CA, USA

ISBN 978-3-031-58444-2 ISBN 978-3-031-58445-9 (eBook)
https://doi.org/10.1007/978-3-031-58445-9

This Springer imprint is published by the registered company Springer Nature Switzerland AG
The registered company address is: Gewerbestrasse 11, 6330 Cham, Switzerland

In the high and pure philosophic desire
to remove error as well as discover truth —

Michael Faraday (1832)

Foreword

The author of this book, Dr. Adrianus Kalmijn, was a true pioneer with a unique understanding of biology and physics. This enabled him to arrive at conclusions, which no one before him had been able to deduce.

Since the first anatomical description of the mysterious tubes in the head of elasmobranchs, today known as Ampullae of Lorenzini, their function was a puzzle. These tubes were demonstrated to be sense organs, and they were proven to have great sensitivity to an array of stimuli. They are sensitive to changes in hydrostatic pressure, salinity, temperature, and electric fields, but for decades the biological significance was very uncertain.

However, a series of very well-planned experiments, best described by Kalmijn (1971), demonstrated that a dogfish could detect the presence of a plaice buried in the sand, only because of the electric fields from the respiratory movements of the operculum of the flatfish. These experiments immediately convinced the conservative and skeptical scientific world of their function.

Later Kalmijn designed experiments, in the open ocean and in tanks, which allowed him to conclude that the extraordinary sensitivity to weak electric fields allowed elasmobranchs to sense their swimming direction based on their movements in the geomagnetic field.

Kalmijn also made very significant contributions to the physics of hair cells in the lateral line and inner ear.

In 1984, I met Vera and Adrianus Kalmijn in San Diego during a research visit to the Scripps Institution of Oceanography. Their hospitality and friendly way to meet other scientists made us friends, and when I returned, with my family, to San Diego in 1986, we collaborated on experiments designed to disclose the sensitivity to hydrostatic pressure of the Thornback Guitarfish (*Platyrhinoidis triseriata*). We found that this ray could detect changes of less than 10 cm hydrostatic pressure in a shallow tank. Unfortunately, we never finished the experiments because I had to return to Denmark. But I was much impressed when I realized how strict Kalmijn was regarding control of the experiments. This was somewhat in contrast to what we see today, where many researchers hurry to publish preliminary results. Kalmijn never jumped to conclusions.

My family and I spent many good times together with Vera and Adrianus. We shall never forget the trips to the Anza Borrego desert, which they introduced to us.

It is a great pleasure for me to recommend this book. While it may be a challenge to read, I am sure that it will be an essential road map for scientists and teachers interested in the biology of elasmobranchs or general animal orientation. It really is a most important text in this field, and we are lucky that Adrianus Kalmijn spent so much time to clarify his thoughts. We also owe much gratitude to Vera Kalmijn and their children who helped move this work to publication.

Aarhus, Denmark
2023

Jørgen Mørup Jørgensen

Preface

Dear reader, you are welcome to delve into the just completed monographic treatise on motional electric orientation in sharks and rays, if you have the courage, but be aware, it was not written with you personally in mind. The author does not know your scientific background nor what you hope to learn from his discourse. You may very well consider it scholarly gibberish and put it aside. Actually, he wrote the article in an endeavor to explain to himself what he has been asserting for half a century about sharks and rays orienting in the oceans' electric and magnetic fields, with the ultimate goal to ascertain whether at his age he really understands the intricacies of his lifetime scientific endeavor. The present version is not intended as a final statement but as a platform for further discussion. Hence, if you find errors or disagree, as you certainly will here and there, or despite great effort cannot follow his concise reasoning, he would greatly appreciate you letting him know, so that together we can discover the truth and lay a firm physical biology foundation for future generations of scientists.

The art of biological science is to ask the animals experimentally the right questions, using the right methods, and properly to interpret their behavioral answers. Certainly, sharks gave the author many invaluable hints. Actually, to grasp their sensory world, he endeavored to view their underwater habitat as if he were a shark himself. As a result, he not only gained insight into the animals' electric and magnetic sense, but also got now and then rather vicious, especially when encountering scientists in his field of research who have not made a serious effort to inform themselves of the rich legacy of our trusted forefathers, Faraday, Maxwell, and Einstein, upon which the present author has founded his hard-wrought contributions to science. With regard to well-versed inquisitive readers, though, he wishes them great success in science, in the hope and expectation that physics will start making sense to shark biologists and for physicists shark biology will become an animated field of research.

Though I wrote the article to the best of my abilities in the crowning years of my career, I sincerely hope that it inspires you, the reader, to further this exciting field of research. However, I shall warn you in advance, the initial highlights, which lay bare the physics bedrock of the motional electric orientation theory, are very concise. Furthermore, the first section is rather grueling with lots of nitty-gritty physical intricacies, many of which have escaped the attention of my predecessors. The second section, on the exploratory movements sharks must make to detect the electric and magnetic fields, is still exacting but may be more appealing biologically. The last section is not nice at all, for which I feel regretful. I wish I would not have had to include it. However, it is the obligation of a scientist not only to state what he considers right, but also to state what he considers wrong and to do so in clear scientific terms. I have tried my very best, even though I realize that my scientific adversaries, who I mention by name on some of the darker pages, may be displeased, for which I apologize.

Perhaps, my earlier writings on the topic are more palatable to start with. Instead of aggravating you with the present harrowing narrative, I merely aspired to relive my line of thoughts concerning the theory of electric and magnetic orientation in sharks and rays as it evolved from reading Faraday's Lecture on magneto-electric induction (1832), perusing von Arx's Physical Oceanography textbook chapter (1962) on towed electrodes to measure the velocity of ocean currents, and scrupulously equation-by-equation analyzing the classic foundational article of Longuett-Higgins, Stern, and Stommel (1954), while I was still a biology and physics student at the University of Utrecht in The Netherlands, followed by years of reflection and experimentation at the Woods Hole Oceanographic Institution, Massachusetts, and at the Scripps Institution of Oceanography, California, leading by good fortune to my present position as an Emeritus with return to active duty, enjoying the use of my Scripps biophysical laboratory and marine research facility. I feel strongly urged to complete my lifetime scientific endeavor by filling the voids I left behind and correcting my mistakes, for the sake of my own fulfillment and in deference to my teachers, our venerable scientific forefathers and, above all, the awe-inspiring sharks and rays.

The article was written and checked in several stages over a long period of time. The main text is interspersed with mini-essays, personal opinions, and bold statements. Some elaborations are put in appendices and a scholium, so as not to interrupt the flow of the discourse. Vital issues are discussed more than once in different contexts, since electric and magnetic misbeliefs have led to widespread persistent but badly-off instances of alternative interpretations. In short, the manuscript offers a mélange of honed utterings, repeatedly hammering on a variety of fundamental anvils. Deplorable enough, the author now and then lashes out in exasperation, feeling overwhelmed by the piles of road clutter along the challenging path to the truth. Rather than being formal and conventional, he often writes in personal terms to pour out his heart and soul in an attempt to break the deadlock in the exciting field of research he initiated. Why should I hide my personal emotions in presenting my scientific data and thoughts behind the stifling reporting constraints? I am not a robot.

Admittedly, I am a lone scientist, scientifically feeling most at ease with my revered teachers of the past, who I yearly meet—in imagination—for a chat on a fallen tree trunk in my holy woods in The Netherlands. From very early on, I enjoyed Galileo's conversational writings about relativity, Newton's hypotheses *non fingo*,

Faraday's magneto-electric transduction of a ship passing on the surface of the water, Maxwell's exacting treatise (1891), Heaviside's genius understanding and blunt criticism, my compatriot Lorentz's namesake force, and Einstein's insights in the nature of physics, especially his 1905 motional electric view. If I unwittingly imitate any people, they are these old-time scientific heroes of mine. Concerning my academic education, I am indebted to Dr. S Dijkgraaf and to Dr. T. H. Bullock, who gave me the opportunity to conduct my Ph.D. research, though—by my fault—I was too physics-minded for further joint efforts with them. I agree with them that I am a difficult person. Nevertheless, we remained trusted friends for all the remaining years. Concerning my tenacious perseverance, I must thank my outspoken adversaries, too many to name.

Unfortunately or not, I have one stubborn drawback in that I am reluctant to write many short articles for the sake of promotion or for securing contracts and grants. Yet, I feel strongly urged to write comprehensive articles when I feel scientifically ready for them, not because of the trite saying "publish or perish." Fortunately, I survived anyhow. Future will tell whether I made the right choice. Beside the present discourse, I have written a companion motion-electric rebuttal manuscript that will be submitted. Furthermore, I have prepared fresh translations of the old Italian and Latin texts on the ampullae of Lorenzini of sharks and rays and on the lateral line of fishes in general. I believe them biologically to be more accurate than earlier translations, having worked on the topics myself for many years. Yet, I will not tell you which further manuscripts I am working on until I am sure that they are ready for publication—deo volente. Life on earth is so long as it is, no longer. Luckily enough, legacies last. Hence, they had better be well-founded.

Since English is not my native language, I do not claim consistently to be correct grammatically and idiomatically. It is not the British English I learned in the Dutch high school either. It is some kind of Californian English without the "real good" instead of "really well." My proud Dutch accent is well known and immediately noticed after I speak three words or even fewer. I have been blamed that my writings are all about me, overly autobiographic. Not true, they all are about what the sharks and rays generously shared with me concerning their keen electric and magnetic sense, behaviorally, electrophysiologically, anatomically, and theoretically. No hearsay, all pure physics and biology conquered by me over the years, based upon the foundations laid by my scientific forefathers. I am merely the fascinated intermediary, seeking to explain what my forefathers discovered not too long ago, yet the venerable sharks and rays have practiced for millions of years.

What do I hope to achieve with the present article? I certainly do not expect all scientists interested in the electric and magnetic sense of sharks and rays to digest my discourse in its ghastly details. It may not be needed either. Yet, I am eagerly looking forward to learning of a few scientifically accomplished readers who publicly dare to concur with the article, if necessary after clarifying some issues and, where needed, correcting some minor or even major mistakes in discussion with me. I hope, these experts will in writing support the hard-wrought conclusions of my research to assure those readers who are not trained to scrutinize the exacting physical and biological particulars of my treatise. Scientists are mortal, but science will evolve forever, however long that may be.

San Diego, CA Adrianus J. Kalmijn

Acknowledgments

My husband, Adrianus J. Kalmijn, devoted his life to understanding and explaining the electric and magnetic orientation in sharks and rays through a deeply rooted and exacting physics approach. He developed this manuscript over several years and we worked together on proofreading and editing it to the last days of his life in 2021. One of Ad's last wishes was that this work would be published and he entrusted me with shepherding the manuscript through the final steps. With profound respect and admiration for Ad's lifelong devotion and contribution to science, my sons, Jelger and Sander Kalmijn, and I prepared this manuscript for publication. I would like to express our gratitude to Scripps Institution of Oceanography at the University of California San Diego, especially to Dr. Douglas Bartlett who has been influential and a great support in getting this monograph published.

Ad received lifelong support from the Office of Naval Research and funding from the W.M. Keck Foundation and the Office of Naval Research for the Electromagnetic Research Facility.

Vera Kalmijn, verakalmijn@gmail.com

Contents

1 Introduction 1
1.1 Highlights 7
1.2 Introduction to Biology and Physics of Motional Electric Orientation 16
1.3 Behavioral Evidence of Electric and Magnetic Orientation in Sharks and Rays 22
References 25

2 Passive and Active Modes of Electro-orientation 27
2.1 Passive Mode of Electro-orientation: Shark Detecting Drift in Ocean Stream 30
2.1.1 Physical Principles of Passive Electro-orientation: Forces on Ionic Charge Carriers 30
2.1.2 Physical Principles of Passive Electro-orientation: In Terms of Voltages 37
2.1.3 Further Developments: Physical Features of Ocean Streams 59
2.1.4 Relaxing the Constraints, Extending Models to Meet Nature 61
2.1.5 Biological Interpretation: Orientation in Ocean Streams 63
2.1.6 Ocean Streams Observed in Frame of Drifting Shark 66
2.2 Active Mode of Electro-orientation: Shark Detecting Compass Heading 68
2.2.1 Physical Principles of Active Electro-orientation: Forces on Ionic Charge Carriers 68
2.2.2 Physical Principles of Active Electro-orientation: In Terms of Voltages 73
2.2.3 Further Developments: Dependence on Resistivity of Medium 83
2.2.4 Relaxing of Magnetic Constraints, Extending Models to Meet Nature 87

2.2.5 Biological Interpretation: Detection of Compass Heading 89
2.2.6 Compass Heading Sensed in Frame of Shark 90
2.3 Additional Remarks 91
2.3.1 Relative Motions Between Body Parts 91
2.3.2 Reactance of Tissues and Excitability of Receptor-Cell Membranes 92
2.3.3 Passive and Active Electro-orientation: Concluding Remarks 94
Appendices 95
Appendix 1. Definition of Symbols in Summary Statements 95
Appendix 2. Mathematical Concepts of Forces and Gradients 97
Appendix 3. Concepts of Potential Difference, Voltage, Divergence, and Curl 98
Appendix 4. Mathematical Consistency Concerning Divergence and Curl 99
Appendix 5. Detailed Analysis of a Technical Example 100
Appendix 6. Axiom: Electric Field Inside and in the Vicinity of a Resistive Sphere 101
References 104

3 Sharks Exploring Ambient Electric and Earth's Magnetic Fields 105
3.1 Exploratory Movements Integral Part of Sensory System 105
3.2 Orienting in DC Ambient Electric and Magnetic Fields with AC Receptors 107
3.3 Magnetic Compass Orientation, Rectilinear Acceleration Procedure 109
3.4 Ambient Electric Field Orientation, Turning Procedure 112
3.5 Magnetic Compass Orientation, Turning Procedure 114
3.6 To Summarize the Various Exploratory Movements 117
3.7 Ambient-Electric and Compass Orientation, a Worked-Out Example 117
3.8 Swerving Motions During Steady Swimming 119
3.9 Motion-Electric Field from Undulating Motions 124
3.10 Generalization of Electric and Magnetic Procedures 124
3.11 Time-Domain or Frequency-Domain Processing of Sensory Data? 126
3.12 Afterthoughts 129
References 129

4 Sensible and Reprehensible Misunderstandings 131
4.1 Before 1974, the Emergence of Electroreception in Sharks and Rays 132
4.2 Definition of Electric and Magnetic Sense 135
4.3 Sharks and Rays Practice Faraday's Physics 138
4.4 Kirschvink's Errors 143

4.5 Did Faraday, Maxwell, Einstein, and the Creator Get It All Wrong? 146
4.6 Stick to the Proven Physical Facts and Principles 152
4.7 Two Review Articles Misrepresenting the Original Literature 155
4.7.1 First Troublesome Review Article 156
4.7.2 Second Troublesome Review Article 165
4.8 Per Se, Direct, and Motional Electric Magnetic Detection 172
4.8.1 Legacy of Paulin 174
4.9 The Literature Steadily Slides Even Steeper Downhill 182
4.10 Concluding Remarks by the Author 189
4.11 Concluding Remarks by Sharks and Rays 191
References 194

Chapter 1
Introduction

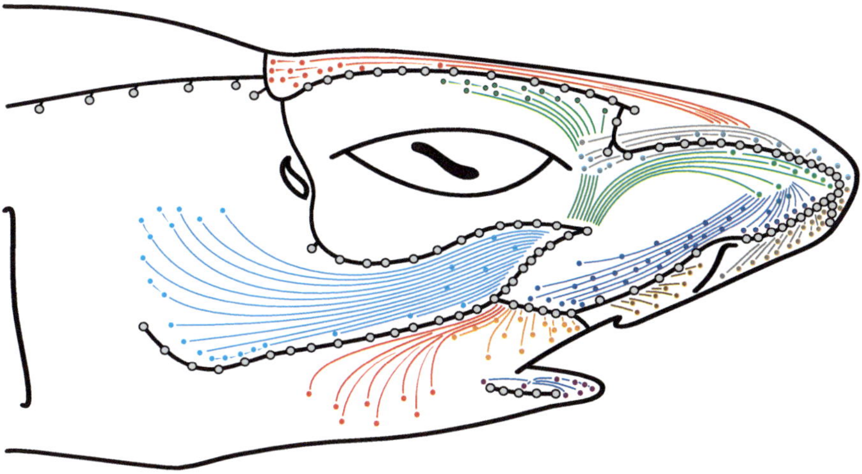

The principal characters Scyliorhinus and Torpedo
Ampullae of Lorenzini and lateral-line canals sensory systems of sharks and rays

In the shark *Scyliorhinus canicula*, the ampullae of Lorenzini are clustered in three paired capsules deep under the skin, the rostral, infra-orbital, and the small mandibular capsule, from where the subcutaneous ampullary canals spread over the head in virtually all directions. The proximal ends of the colored traces indicate the positions of the capsules from which the canals originate. The distal ends of the traces indicate where the ampullary canals open to the outside through the skin pores, the solid dots, densely populating the head of the animal. The small circles are the pores of the subepidermal lateral-line canals, in black. Also indicated in black are the naris, mouth, eye, spiracle, and the first gill slit.

A. J. Kalmijn, *Theory of Electric and Magnetic Orientation in Sharks and Rays Revisited*, https://doi.org/10.1007/978-3-031-58445-9_1

Differential pair of ampullae proper, ampullary canals, skin pores, and afferent nerve fibers

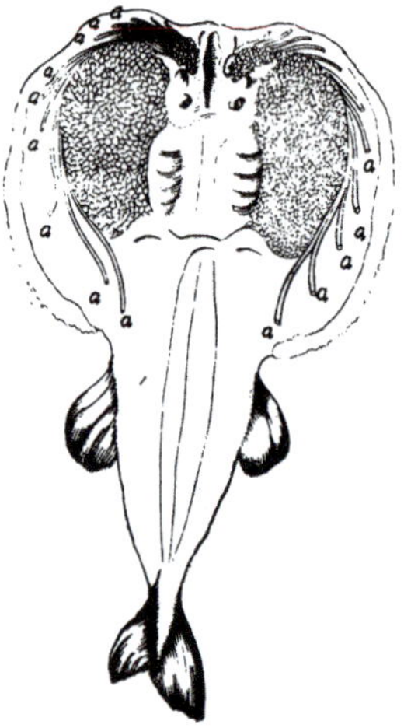

aaaaa, ec. Sono i canali fin quì descritti, che terminano nella pelle.

bb, Luogo dove i canali suddetti con l'altra estremità vanno ad unirsi a i globi di sopra descritti.

Figure four of Table one, in which: aaaaa, etc. are the *canali* described thus far, which terminate in the skin. bb, the place where the above-mentioned *canali* at the other end join the globes described above.

Stefano Lorenzini (1678). Osservazioni intorno alle Torpedini

Pictorial of fundamental observations

Bottom dwelling dogfish *Scyliorhinus canicula* orienting in bioelectric field of prey

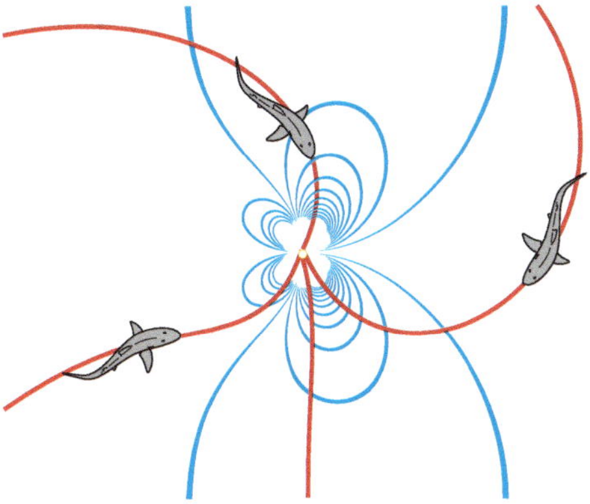

Guided by approach algorithm sharks zero in on quadrupole source simulating prey

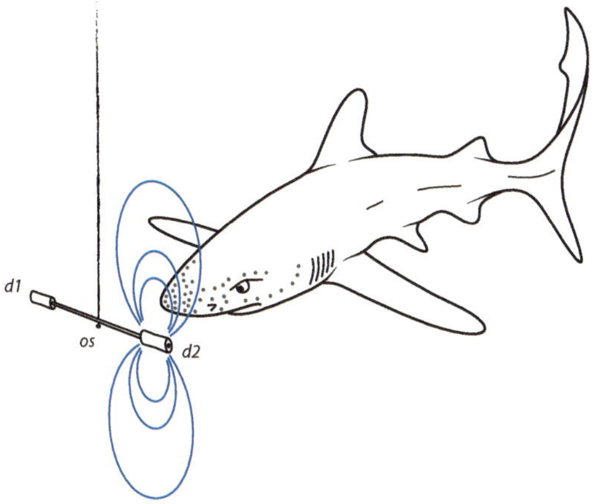

Pelagic blue shark *Prionace glauca,* roused by odor, attacks prey simulating dipole source

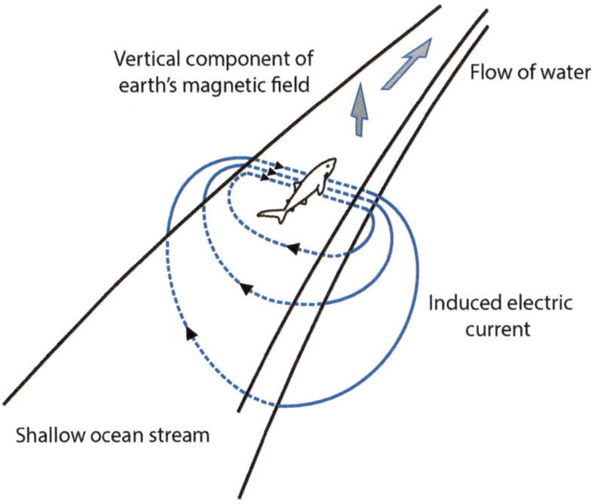

Shark electrically sensing drift in ocean stream

$$\nabla \varphi = \mathbf{v} \times \mathbf{B} - \rho \mathbf{J}$$

Equation governing motional electric orientation

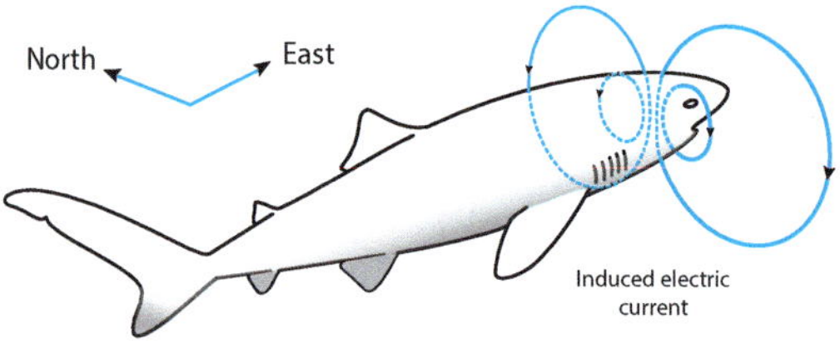

Shark electrically sensing magnetic compass heading

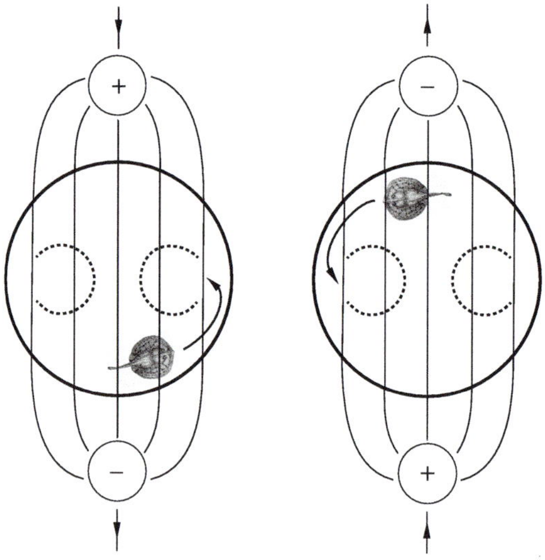

Stingray *Urolophus halleri* seeks food in enclosure to the left of ambient electric field

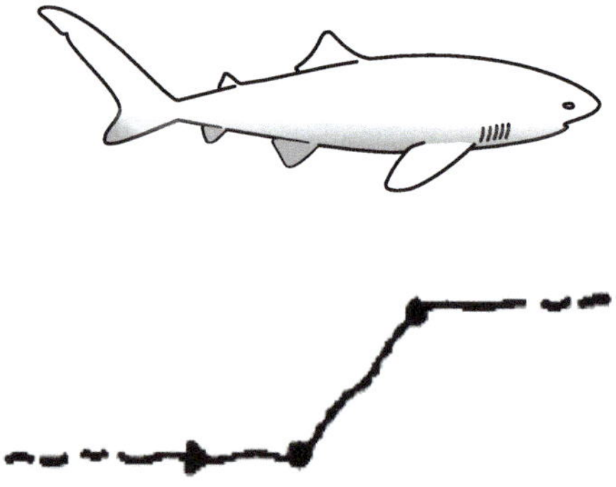

At sea, shark follows 60-degree rotation of ambient electric field

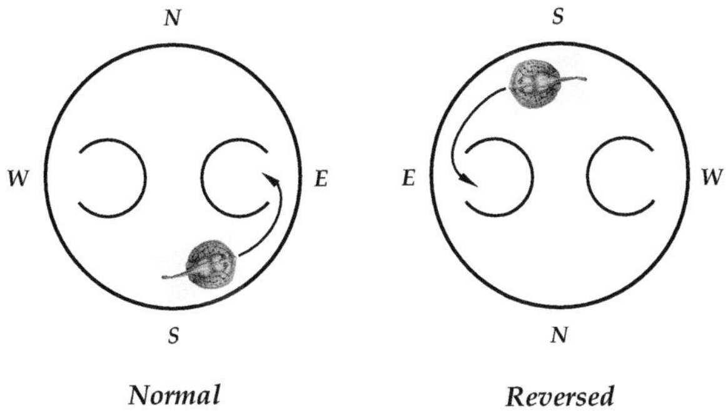

Stingray *Urolophus halleri* seeks food in enclosure in the East of earth's magnetic field

EMRF and RV Boston Whaler

Electromagnetic Research Facility. 3-axis, 18-coil ambient *magnetic* field control system surrounding cylindrical, 12 ft in diameter, 7 ft deep seawater tank with ambient *electric* field control system. Modular building and EM control systems designed by the Principal Investigator, Dr. Ad. J. Kalmijn.

Project's Research Vessel RV Boston Whaler on Bimini Bay for Orientation Experiments

The founding fathers

Michael Faraday

James Clerk Maxwell

Oliver Heaviside

Hendrik A Lorentz

Albert Einstein

1.1 Highlights

Fundamental principles ascertained by Faraday

The first four paragraphs are concise statements concerning the physical foundation of electric and magnetic orientation in sharks and rays. They are not meant to be understood at first glance unless the reader is more sharp-witted than the author, who has struggled with the physical intricacies for at least half a century. Nevertheless, after the persevering reader has digested the entire manuscript, the first four paragraphs should appear sensible and ready to be commented on to improve and, where needed, to correct them. The descriptive mathematical-physics symbols used are defined in the text and, for convenience, further clarified in Appendix 1—Chap. 2. *The superscript refers to the region where the physical quantity is measured, and the subscript to the region from which the physical quantity originates.* Figures 2.1 through 2.6 and their legends, to be discussed in detail at a later stage, may be also consulted by the reader who desires a pictorial illustration of the physical relations.

- A *wind-driven, open-ocean stream* flowing at a velocity $\mathbf{v}$ is—due to interaction with the *vertical* component of the earth's magnetic field $\mathbf{B}_v$—subject to a *horizontal*, stream-wide motionally induced electric field $\mathbf{v} \times \mathbf{B}_v$. In the stream, the ensuing induced electric current of density ${}^{str}\mathbf{J}_{str}$ causes an ohmic resistive field $-\rho^{str}\mathbf{J}_{str}$ and, by accumulating − and + charges at the lateral boundaries of the stream, an electrostatic motional electric field ${}^{str}\mathbf{E}_{str}$. The $-\rho^{str}\mathbf{J}_{str}$ and ${}^{str}\mathbf{E}_{str}$ oppose the $\mathbf{v} \times \mathbf{B}_v$. The electric field of the − and + charge accumulations furthermore drives a return current of density ${}^{stw}\mathbf{J}_{str}$ through the still deeper water relative to which the stream is flowing. Voltage sensing electrodes drifting with the stream measure the $-\rho^{str}\mathbf{J}_{str}$ resulting from the induced electric current ${}^{str}\mathbf{J}_{str}$, where $-\rho^{str}\mathbf{J}_{str}$ equals $-\mathbf{v} \times \mathbf{B}_v$ *minus* the ${}^{str}\mathbf{E}_{str}$ of the charges at the boundaries of the stream. In a shallow, by the return current heavily loaded stream, the $-\rho^{str}\mathbf{J}_{str}$ largely overrides the ${}^{str}\mathbf{E}_{str}$, thus offering a close estimate of $-\mathbf{v} \times \mathbf{B}_v$, hence of the velocity $\mathbf{v}$ of the stream. Deeper, electrically less-loaded streams accrue larger charge accumulations at their lateral boundaries, causing ${}^{str}\mathbf{E}_{str}$ to gain in strength at the cost of $-\rho^{str}\mathbf{J}_{str}$, leaving for the electrodes less of the velocity $\mathbf{v}$ of the stream to measure.
- A *shark drifting in a wind-driven, open-ocean stream* flowing at a velocity $\mathbf{v}$ is—due to interaction with the *vertical* component of the earth's magnetic field $\mathbf{B}_v$—subject to a *horizontal* motionally induced electric field $\mathbf{v} \times \mathbf{B}_v$, *and* to the motional electric field ${}^{anl}\mathbf{E}_{str}$ of the − and + charge accumulations at the boundaries of the stream. Since the animal is more resistive than the surrounding seawater, the electric current induced in the stream accumulates + and − charges on opposite skin surfaces of the animal, causing part of the current to flow around, instead of through the animal. In the stream, the electric field ${}^{str}\mathbf{E}_{anl}$ of the charges on the shark's skin surfaces partially offsets the electric field ${}^{str}\mathbf{E}_{str}$ of the charges at the stream's boundaries. Since, at constant $\mathbf{v} \times \mathbf{B}_v$, the sum of ${}^{str}\mathbf{E}_{str}$ and $-\rho^{str}\mathbf{J}_{str}$ remains the same, the decrease ${}^{str}\mathbf{E}_{anl}$ causes an increase $-\rho^{str}\mathbf{J}_{anl}$. Hence, the stim-

ulus the shark detects is $-\rho^{str}\mathbf{J}_{str}$ *plus* $-\rho^{str}\mathbf{J}_{anl}$ integrated across the body along a seawater path, which equals $-\rho^{anl}\mathbf{J}_{str}$ *plus* $-\rho^{anl}\mathbf{J}_{anl}$ integrated along a path through the animal. In a heavily loaded stream, the shark may *infer* from the $-\rho^{anl}\mathbf{J}_{str} - \rho^{anl}\mathbf{J}_{anl}$ the upstream direction and the speed of *its drift*. Upon entering freshwater, fewer or opposite charges accumulate on the skin, reducing the effective stimulus.

- A *shark traveling in still ocean waters* at a velocity **v** is—due to interaction with the *horizontal* component of the earth's magnetic field $\mathbf{B}_h$—subject to a *vertical*, body-high, motionally induced electric field $\mathbf{v} \times \mathbf{B}_h$. In the animal, the ensuing induced electrical current of density ${}^{anl}\mathbf{J}_{anl}$ causes an ohmic resistive field $-\rho^{anl}\mathbf{J}_{anl}$ and, by accumulating − and + charges at its ventral and dorsal skin surfaces, a motional electric field ${}^{anl}\mathbf{E}_{anl}$. The $-\rho^{anl}\mathbf{J}_{anl}$ and ${}^{anl}\mathbf{E}_{anl}$ oppose the $\mathbf{v} \times \mathbf{B}_h$. The electric field of the − and + charges also drives a return current of density ${}^{stw}\mathbf{J}_{anl}$ through the seawater relative to which the shark is moving. In the traveling shark, the ventro-dorsal line integral of $-\rho^{anl}\mathbf{J}_{anl}$ across the body is the stimulus the animal detects. Since the animal is heavily loaded by the return current, $-\rho^{anl}\mathbf{J}_{anl}$ prevails over ${}^{anl}\mathbf{E}_{anl}$. Hence, $-\rho^{anl}\mathbf{J}_{anl}$, which equals $-\mathbf{v} \times \mathbf{B}_v$ *minus* ${}^{anl}\mathbf{E}_{anl}$, approaches $-\mathbf{v} \times \mathbf{B}_h$, though does *not* equal $-\mathbf{v} \times \mathbf{B}_h$. The shark may *infer* from $-\rho^{anl}\mathbf{J}_{anl}$ *its magnetic compass heading*. Since sharks entering brackish waters, freshwater fish, and air-borne birds are less to hardly electrically loaded by their surrounding media, sizable amounts of charges accumulate on opposite skin surfaces, the ${}^{anl}\mathbf{E}_{anl}$ of which causes the effective stimulus $-\rho^{anl}\mathbf{J}_{anl}$ to range from weak to virtually nil.
- To complement the situations considered thus far, a shark *drifting in a wind-driven, open-ocean stream* flowing at a velocity **v** is—due to interaction with the *horizontal* component of the earth's magnetic field $\mathbf{B}_h$—subject to a *vertical*, stream-high, motionally induced electric field $\mathbf{v} \times \mathbf{B}_h$, in this instance depending in strength and polarity also on the stream's magnetic compass heading. However, as a result of the electrically insulating seawater-air interface, significant − and + ionic charges accumulate at the surface and the bottom of the shallow stream over most of its width. Hence, vertically, the ${}^{str}\mathbf{E}_{str}$ prevails over the $-\rho^{str}\mathbf{J}_{str}$, *leaving little for the shark to detect* ventro-dorsally. Furthermore, a shark *traveling* in the ocean at a velocity **v** is—due to interaction with the *vertical* component of the earth's magnetic field $\mathbf{B}_v$—also subject to a *horizontal*, body-wide, motionally induced electric field $\mathbf{v} \times \mathbf{B}_v$. Since in the ocean the shark is heavily loaded electrically by the seawater medium, the charge accumulations on its left and right skin surfaces are small and so is the ${}^{anl}\mathbf{E}_{anl}$, allowing $-\rho^{anl}\mathbf{J}_{anl}$ to reach nearly its full strength, thereby providing the vertical magnetic information needed for sharks to *sense the inclination of the earth's magnetic field*, hence their magnetic latitude.
- Besides the motional electric fields signaling drift and magnetic heading, many other large-scale and more local electric fields exist in the oceans from which sharks may derive vital information. The electric return currents of ocean streams invade adjacent and deeper waters, even penetrate sandy ocean bottoms. The fluctuating fields of tidal flows, the fields induced by solar-driven variations in the earth's magnetic field, the telluric fields stemming from the earth crust, the

electrochemical fields of contiguous bodies of seawater and bottom sediments are all widespread occurrences, and so are, despicably enough, the galvanic fields of oil rigs, wind farms, and passing ships, electrically polluting ocean waters. Interspersed with the local bioelectric fields of aquatic animals and the image fields of resistive objects excited by the ambient electric fields, the large variety of fields creates an electrically diverse subsea world. Lastly, the electric fields of the water motions in the surf and shallow coastal waters offer a wealth of sensory cues. Interpretation of the diverse electric fields requires detection of their spatial distribution and time course, and above all great expertise on the part of the recipient animals, evidently gained during the process of evolution and from lifetime experience.

Fundamental principles practiced by sharks and rays

- The guiding principle of the present discourse is the observation that not only the electric fields sharks are subject to in ocean streams but also the electric fields they themselves induce by moving in the earth's magnetic field are received by the animals as imposed uniform $-\rho\mathbf{J}$ fields. That is particularly evident in the shark's frame of reference, in which neither the animal nor the sense organs are moving, hence $\mathbf{v} \times \mathbf{B}$ is nil, as pointed out by Einstein (1905). Yet, the two kinds of uniform $-\rho\mathbf{J}$ fields are readily distinguished from one another as they fundamentally differ in their dependence on the shark's exploratory movements, which render them unambiguously discernable as signaling either drift or magnetic compass heading. To sense the horizontal component of ambient *electric* fields, sharks may turn left or right at arbitrary but constant forward speed. Then the horizontal ampullary pairs stimulated least point in the direction of the field, whereas the horizontal ampullary pairs stimulated most are oriented normal to the field. To sense the horizontal component of the earth's *magnetic* field, sharks may accelerate rectilinearly in two different directions. The stimuli thus received by the ventro-dorsal ampullary pairs inform the animals of their magnetic headings.
- Be well aware, in the present article, I do not intend to tell the reader how sharks and rays orient electrically and magnetically. I merely want to convey how I imagine that they may do so on the basis of what I learned from the structure and physiology of the ampullary system, from the animals' behavior in the laboratory and at sea, and from the established physics governing the oceans' electric and magnetic fields. I started out conceiving of the most elementary linear and angular accelerations sharks may execute to detect the oceans' electric and magnetic fields, and how the animals may derive the desired information from them in conformance with the laws of physics. Based on the results, I examined the moves sharks make in daily life, singly or in smooth succession, to make sure that my line of reasoning was behaviorally realistic, even though sharks may not operate as I imagined. Hence, my studies aim to foresee the exploratory movements consistent with the animals' electrically and magnetically guided orientation behavior. To establish the validity of my inferences, we shall actually measure the motion-electric fields and ascertain under biologically valid conditions whether sharks and rays behaviorally indeed operate as envisioned, a task I may have to leave to my successors.

- The motional electric signals sharks receive from the oceans' ambient electric field and from the earth's magnetic field are at constant velocity inherently of a DC nature, whereas—seemingly in conflict—the sharks' electroreceptors, the ampullae of Lorenzini, are not DC but low-frequency AC sensors. Even so, the oceans' electric and magnetic fields the animals day and night depend on are predominantly quasi-steady stationary fields. This dilemma has baffled many shark experts, some of whom even ventured to reject my original physical biology theory of electric and magnetic orientation on these grounds. Yet, the resolution of this non-issue is straightforward. As I emphasized in my very first 1974 version of the theory, the ampullae of Lorenzini are part of a comprehensive sensory system, comprising not only the peripheral sensors, but also the central nervous system, the behavioral feedback, the shape and constitution of the animal's body, and the seawater medium (Kalmijn 1974). Just as sharks must make the proper movements to distinguish between the different kinds of electric fields, they must make the proper movements to transfer the DC stimuli into the low-frequency AC range of the ampullary sense organs, movements that are all part of the animals' natural behavior.
- When I state that sharks may *infer*—what we call—their drift in ocean streams or may *infer*—what we call—their magnetic heading, I merely intend to express that the electrical information the animals receive in daily life would allow them to do so, at least, if they would have the cerebral faculties of human beings. We would be able to, at least, if we would have the sensory abilities of the animals. The present article concludes that sharks indeed receive all the information needed to take full advantage of the orientational cues offered by the electric fields of ocean streams and—motion-electrically—by the earth's magnetic field, but it does not infer that they interpret them in terms of *drift* and *magnetic heading*. For sharks, there are simply two kinds of ambient electric fields. The one they may use to set out and maintain a direction in space, which we call their magnetic heading. The other they may use to correct that direction for what we call their drift with the flow of water. Just imagine a shark using the two fields in crossing the English Channel from Dover to Calais, correcting for the strong sideways tidal flows. They need not know of ocean streams and magnetic fields. They must only know how to use the electrical signals they receive to arrive at their destination.

From prey detection to orientation in the ambient electric and earth's magnetic fields

- The notion of detecting the oceans' ambient electric fields and, initially more conjecturally, of detecting the earth's magnetic field motional-electrically may have passed the minds of earlier-day scientists, though without having the correct physical insight and not knowing of a sufficiently high electrical sensitivity in marine animals. First, a conclusive instance of electroreception without the use of electric organs was needed to establish beyond doubt the very existence of a sensory system so foreign to us. Thus, the results of my pioneering studies (Kalmijn 1971) on the electrical detection of prey by sharks and rays provided a well-established starting point for a thorough exploration of the electric and magnetic undersea world of sharks and rays, proving the animals to detect electric fields down to the nanovolt/cm level, amply meeting the requirements for the theory of motional electric and magnetic orientation. Interestingly, the orientation in ambient electric and magnetic fields are, in my opinion, based upon the same algorithm as the electrically guided approaches to prey items, to wit, by keeping an initial set angle relative to the electric field lines or, equivalently, to the isopotential planes. The lateral-line and inner-ear senses apparently apply this orientation algorithm as well.

- To gain further insight into the electric and magnetic sensory world of sharks and rays, first of all, one wonders which are the physical features of the fields the animals encounter at sea and what information do they offer? Next, the question arises whether and how the animals make biologically significant use of the information. To learn about the fields, we—as oceanographers—conduct physical measurements. To ascertain whether the animals use the electric and magnetic data for the causes we envision, we—as biologists—conduct behavioral experiments. As to how the animals interpret the sensory data, they surely do not follow our elaborate scientific reasoning. We need not bestow them with great intellect. They rely on innate and gained experience. For them, it is not hard science but purely an effort to thrive and survive by making the best use of the limited information they receive. Naturally, animals rely on their entire complement of sensory modalities. In sharks and rays, the inner-ear and lateral-line sense organs may serve particularly important roles in controlling their rectilinear and angular exploratory movements as part of their electrosensing. Actually, our own behavioral performances also rely more on lifetime experience than on academic wisdom.
- Traveling sharks are alerted to the presence of local sources of electric fields, such as prey, obstacles, and boundaries by unforeseen changes in the strength and direction of the fields they rely on to sense their drift and compass headings, changes caused by the fields objects produce or by the distortions they create in the ambient electric field, or both. When hungry, especially when motivated by the scent of food, sharks strike at their prey in fast, well-directed moves. During their approaches, they are guided, foremost, by the bioelectric fields of their targets. However, this requires them to take into account the electrical signals they receive from carrying out their attacks in the presence of the ambient electric and the earth's magnetic fields. Hence, we feel logically compelled to assume that sharks continuously monitor the ambient electric field as well as the earth's magnetic field, not only to orient in space, but also to take them into account when homing in on the prey's bioelectric field. My shark-inspired autonomous subsea robotic devices, replicating the electrically guided predatory attacks in the presence of ambient electric fields, corroborate the physical feasibility of concurrently processing the ambient electric and the bioelectric prey field data.
- The detection of ambient electric fields and the detection of the earth's magnetic field are closely related, not just physically by the motion-electric principle, but also biologically in the process of navigation. To travel in an intended magnetic

direction while drifting sideways, a shark must swim in a slightly different direction. It may maintain the corrected direction relying either on the ambient electric or the earth's magnetic field or both, whichever are most readily detected. The magnetic field requires rectilinear accelerations in different directions; the ambient electric field requires turning movements. Physical principles reveal that a head-swaying traveling shark receives very clear signals from the ambient electric field but less pronounced signals from the earth's magnetic field. Hence, by occasionally making the proper exploratory movements to detect the earth's magnetic field, a shark may well calibrate its intended *magnetic* heading in terms of the existing ambient *electric* field, especially in or in the vicinity of ocean streams. In support of the idea, I have observed sharks and rays in Bimini waters along the Gulf Stream to turn left and right in reaction to an electric field which was turned instrumentally, even though the magnetic field remained unchanged.

Biology & physics

- In studying the electric and magnetic sense of sharks and rays, physicists tend in the first instance to describe the animals' underwater world from the perspective of an earth-bound observer. However, to understand the challenges the animals face, we must perceive their underwater world as though through the animals' senses. When we say, "the animal drifts with the flow of an ocean stream or travels relative to the source of the earth's magnetic field," we must realize that in the shark's frame of reference, in which the sense organs are at rest, it is not the animal, but the outside world that is moving. You are reading this sitting still at your desk while the sun sets, aren't you? Already four centuries ago, Galileo (1632) considered relativity as it applies to travelers conducting experiments below deck in a smoothly moving ship. To imagine ourselves to be the ones who are moving, we must mentally transform the sensory information from our own frame of reference to the earth frame. Whether the animals do so, we do not and probably cannot know. Actually, by contemplating the issue, I have consciously come to see the environment turn when I negotiate a hallway corner, though I am aware that it is merely the view of the world from within my frame of reference.
- By the process of evolution, sharks and rays have acquired great skill in applying intricate physics. No wonder, the theory of motional electric orientation has daunted many inquisitive biology-oriented natural scientists and, surprisingly, some overly outspoken physics-trained scientists as well. Although with great trepidation, I feel obliged to point out their failings in order to clarify the issues. However, it is not a matter of my opinion versus theirs but of their opinion being in direct conflict with well-established scientific knowledge, gathered by our venerable forefathers whom we all trust, in particular my teachers Galileo (1632), Faraday (1832), Maxwell (1891), Heaviside (1893), Lorentz (1892), and Einstein (1905). Nearly all the errors made pertain to the underlying physics, though some result from mistaken biological data and ideas, many of which have since been corrected. To pave the way for unhampered scientific progress, it is our moral duty to clear the roadside clutter we and our predecessors have left behind, however unsettling our task may be. I have learned from my adversaries not to be

timid, though it has not made me particularly beloved. Yet, with the present article, I hope to entice daring young scientists to participate in the fascinating study of the oceanic physics practiced by marine sharks and rays.

- It must have taken sharks many years to evolve their intricate motional electric orientation procedures, but why did it take me, the author, so long fully to understand their governing physics, if I do? Surprisingly, several hard-core physicists have wondered about my wet physics approach to animal biology. It certainly is not standard textbook knowledge, though all those I had a chance to reason with agreed in hindsight. Yes, physics applies to wet, ionic animals just as well as to dry, electronic gadgets. However, I studied physics in support of my biology degree to explain the sensory behavior of animals not just to physicists but, in the first instance, to biologists with whom I share my great fascination of animals. It is, therefore, my aim to make biologists—including myself—familiar with the mathematical physics without which the animals' behavioral performances cannot be understood. The best I can do is to discuss with you, the reader, the issues I have struggled with for years. Hence, forgive me, if I seem to belittle you. If I do, I am actually belittling myself by admitting my limited grasp of the animals' sensory systems and explaining why it took me so long to get the present treatise ready for publication. At least I tried hard to come to that point in time.
- Since the prevailing electric fields in open-ocean waters result from large-scale seawater flows interacting with the earth's magnetic field, the sharks' reliance on the oceans' electric fields for the detection of drift is quite sensible. Yet, sharks may also make use of the oceans' motional electric fields for other orientational purposes than detecting drift, especially where the fields invade adjacent still waters. It is the task of the animals to interpret the nature and biological significance of the various electric fields they detect. In the present article, we will focus on the detection of drift side by side with the detection of the earth's magnetic field, both of which are founded upon the same motional electric principles and in open-ocean navigation play complementary roles. However, whereas the information the magnetic field offers is straightforward, the information the electric fields of ocean streams provide for the detection of drift is less evident, as other factors play critical roles as well, in particular the dependence on the electrical loading by deeper water and the ocean bottom. Hence, calibration in terms of drift velocity is a difficult task, requiring further information. However, this is not an issue when the animals merely seek the direction and the greatest velocity of their drift.

Skepticism, doubt, and my search for the truth

- One wonders, is it reasonable to infer that the ampullae of Lorenzini provide sharks and rays, not only with an electric, but also with a magnetic sense? The electrical detection of prey and ocean currents is readily understood, but the earth's magnetic field is something else. Yes, for us it is, but not for the sharks and rays. They motion-electrically perceive the magnetic field just as another ambient electric field, except that it responds very differently to their characteristic, seemingly erratic exploratory movements. By daily experience, sharks and rays appear thoroughly familiar with the two kinds of readily detectable electric fields, but we scientists need a degree in biology and physics to appreciate the animals' electric and magnetic sensory world. But why is there so much doubt and disbelieve in the animals' motion-electric magnetic sensitivity? Is it hard to believe that the animals can properly interpret the motional electric fields? Or are the behavioral tests I conducted to confirm their magnetic orientation deficient? Sure, that would solve all lingering quandaries. But, would we have peace believing that the creator gave sharks and rays an electric sense but failed to make it sensitive enough to detect the biologically most vital electric and magnetic fields in ocean waters?

- A novel new scientific theory usually starts from an inkling based on limited observations. The inkling may be mistaken but driven by excitement, the originator may keep trying hard to prove it right. Often though, the new theory is nonetheless met with disbelieve or outright rejection. That creates an uneasy dilemma. What is the solution? First and foremost, the strategy for defending or rejecting a new theory is to prove it either consistent or inconsistent with established biological and physical knowledge. Therefore, both sides must have the proper biological and physical expertise. Using authoritarian expressions of doubt shall be banned from the scientific literature. Moreover, I consider referring to selected review articles instead of consulting the original literature a scientific offense. As to the ongoing disputes, the main obstacle is the fact that biologists as well as physicists taking part in the field of research do not have the physics proficiency to judge how sharks and rays may use the motional electric fields of ocean currents and those they themselves induce to orient at sea. Ultimately, to reach a definite solution, decisive behavioral experiments shall be conducted, adhering to the strictest rules of physics and under truly valid biological conditions.
- All said and done, but why did I persevere in pursuing a research topic that received so much critique and wonder? What drove me? When I was a graduate student, sharks and rays revealed to me that, without doubt, they respond behaviorally to DC and low-frequency AC nanovolt/cm electric fields, inspiring me to ascertain the biological significance of such an unfamiliar sensory modality. Curious as to the kinds of electric fields sharks and rays may encounter at sea, I not only measured the bioelectric fields aquatic animals produce in the surrounding water but also showed behaviorally that sharks and rays are accurately guided by these fields to their prey, proving them to have a genuine electric sense (Kalmijn 1971). That raised the exciting question whether the animals may also detect the oceans' ambient electric and magnetic fields that, perhaps, may enable them to detect their drift in ocean streams and, motion-electrically, to sense their magnetic compass headings. After 10 years of theoretical studies and behavioral experiments in Utrecht, Woods Hole, and La Jolla, the answer was a definite "yes." That has impelled me to devote my time to learning how they do so physically and biologically. How far I have come after half a century, well the present treatise will tell you.
- The present discourse is the result of my decades long development of the theory of electric and magnetic orientation based on the principle of motion-electric

induction. It is not intended as a review but as an expanded restatement and a desperate effort to eradicate misconceptions. Where readers had difficulty following me, I have elaborated. Where I felt uncertain, I have inquired into the issues more thoroughly. Where I went wrong, I have corrected myself. Where I believe others went wrong, I have, with due regard, tried to set them right, with the risk that I was the one in error. In the end, I hope that what I have contributed to science in my lifetime is in essence true and will remain at the heart of future profound insights. My successors may improve and extend my discourse but, hopefully, not reject it offhand in favor of their own fuzzy inklings. Yet, do not hesitate to call me wrong here or there or everywhere, but be sure and explicit. I have tried to the best of my abilities to recount my research on the electric sense of sharks and rays, inspired by the amicable pioneering approach of Faraday (1832), but also driven by Heaviside (1893) to lash out where called for. Yet, let us scientists be humble. The truth rests with the sharks. The ultimate credit goes to the Creator. Who are we?

Relations to other sensory modalities and to brain processing

- Although the present article focuses on the electric and magnetic orientation of sharks, their electroreceptive system also serves a great diversity of other functions. Well established are the electrically guided approaches to prey items, the detection of conspecifics and animals in general, the responses to electrogenic objects and objects that distort the ambient electric field, the avoidance of aversive man-made electric and magnetic fields, and so forth. Furthermore, sharks have other acute sensory faculties. They are notorious for their good sense of smell. Their eyesight seems well developed, too. In particular, their lateral-line and inner-ear sense organs are highly advanced and by design well adapted to the aquatic environment. Interestingly, their sense of hearing rests on the detection of acoustic particle accelerations instead of acoustic pressure variations (Kalmijn 1988b). The full power of the sharks' electric and magnetic sense relies on the cooperation of the entire complement of their various sensory systems, a topic in dire need of further study. The electric and magnetic sense shows us the importance of understanding the physical principles underlying the various sensory modalities, especially in species living in media of very different conductivities and densities.
- Is the electric sense of sharks and rays really so unfamiliar to us? Evidently yes with respect to sensing the environment, since the aerial medium is not ionic, as seawater and living creatures in essence are. Nevertheless, the principles of electroreception are merely specializations of the basic biophysical principles of animal life, which are not limited to internal signaling and processing, but in the instance of electroreception also are turned to the outside world. In the author's opinion, the electric fields received by sharks and rays excite a negative resistance transduction circuit operating very much like the central nervous dendrites in the analog subthreshold domain, that is, below the level of generating all-or-none action potentials. If my expectation holds, which remains to be ascertained, the ampullae of Lorenzini are of similar significance as the squid giant fibers that

enabled early neurophysiologists to analyze the process of triggering and generating action potentials. The electric sense, though, appears to be of the greatest relevance to the operation of subthreshold graded potentials, the kind of potentials underlying the "thinking" of the brain. This and other exciting biophysical aspects of the elasmobranch's electroreception I hope to come back to later.

The electric sense and the Office of Naval Research

- In the years before 1970, I grew up as a Dutch native to become a natural scientist in my home country, The Netherlands. However, after earning a dual degree in biology and physics, I came to be a Dutch scientist in the United States, to my own surprise. How did that happen? When invited as a visiting scientist by Scripps Institution of Oceanography in La Jolla, California, I fortuitously met at an ONR site review with two Program Managers, who asked me "who are you, what are you working on?" When I told them that I was from The Netherlands, visiting for 2 years, and as a graduate student had found that sharks and rays have an electric sense, they exclaimed, "oh, is that you!" Two weeks later, they formally asked me to join their oceanography research program. This was still in the golden years that ONR funded basic research that their own laboratories may benefit from without asking the scientists what it is good for, which they preferred to decide themselves in the early days. As a result, I have for decades conducted the research reported in the present manuscript under ONR's auspices. I have enjoyed several other generous funding offers, but it was ONR that initiated my lifetime effort in the USA, which I am very thankful for. Yet, my name and accent still attest to my Dutch origin, which I am proud of.

1.2 Introduction to Biology and Physics of Motional Electric Orientation

Biological aspects

The marine habitat offers sharks and rays a great wealth of electric and magnetic information. The worldwide system of wind-driven ocean circulations interacts with the earth's magnetic field creating a global pattern of ambient electric fields. By the same motional electric principle, sharks traveling ocean waters create electric fields informing them of their magnetic heading and the latitude of their position on the earth globe. Animals, plants, and many inanimate objects produce electric fields and all distort ambient electric fields, enabling sharks and the kindred rays to strike at their prey, to avoid obstacles, and to negotiate boundaries. These broad-ranging inferences, inspired by Faraday's "Experimental Researches in Electricity," are founded on the results of the present author's behavioral experiments, conducted over a period of half a century in his marine facilities and in the wild at sea, proving sharks and rays to rely on their acute electrical sensitivity in locating animate and inanimate objects and in orienting in the oceans' ambient electric and magnetic fields. Based on this wide range of behavioral experiments, he felt justified to conclude that sharks and rays have a genuine electric *and* magnetic sense of which the ampullae of Lorenzini are the receiving sense organs.

For a single sensory system to detect electric and magnetic fields for such diverse causes as orientation and object detection seems to require truly sophisticated signal processing capabilities. The challenge is even more dire, as sharks predominantly rely on DC electric and magnetic fields, whereas their electroreceptors, the ampullae of Lorenzini, are not true DC, but low-frequency AC devices, detecting frequencies, say, from 1/16 to 4 Hz. This becomes intelligible when we recognize the sharks' seemingly erratic pattern of moving about in ocean waters as an essential element of the functioning of their electric sense. Their frequent accelerating and turning movements transfer the DC fields into the low-frequency range of their electroreceptors and enable them to stay at all times informed of the ambient electric and magnetic fields, not only for the purpose of spatial orientation, but also to take the signals they receive from the ambient electric and magnetic fields into account when they are electrically guided to their prey or manage obstacles and boundaries. Naturally, other sensory modalities play vital roles as well, in particular the sense of smell and the hydrodynamic and inertial senses of the lateral line and inner ear in eliciting and controlling the exploratory movements.

Foremost questions are: which, in the presence of the earth's magnetic field, are the electric fields a shark receives when passively drifting in an ocean stream and when actively moving with respect to the water, or both? And, in object detection, which are the electric fields a shark receives when approaching prey or avoiding obstacles? To study the shark's electric and magnetic sensory world, we traditionally start from the sources that produce the fields. That is the *forward problem*. However, concerning the fields of objects, the animal faces the *inverse problem*: where do the signals it detects originate from and what do they reveal about the sources? Since sharks detect only a small part of the electric field of a distant source, namely the part acting directly on the electrosensitive skin surface, they do not obtain complete information as to the nature and position of the source, whether live prey or a nearby obstacle. Hence, from the limited data sharks receive, they must infer a great deal to lessen the uncertainty. In prey detection, sharks may even not electrically know in advance the position of the source, but let themselves be guided by the multipole nature of the field to reach their target.

One may wonder how sharks sort out the very different electrical signals they receive: from underwater objects enabling them to zero in on their prey, from ocean streams informing them about their drift, from their own motions in the earth's magnetic field giving them their compass heading, and from all other electrical sources, we either know or have not yet thought of. Prey detection seems simple enough, but on close inspection it is not. Prey items are electrically not like point sources rectilinearly radiating in all directions. The prey fields instead arise from arbitrary multipole sources, the field lines of which are not, in general, linearly directed at the point of origin. Conversely, the orientation in ocean streams and in the earth's magnetic field is straightforward, though the governing physics is rather demanding. Still, whereas my original discovery of the electrically guided attacks at prey fish has never been challenged, it is often said, with some mistrust, that "scientists even believe that sharks and rays orient in the ocean's electric and magnetic fields," as if it were less plausible.

Since my student pioneering research on the electric and magnetic sense of sharks and rays, I have gradually gained a profound insight into the animals' multifunction electrosensory repertoire. Admittedly, several of my ideas concerning critical issues remained tentative, some for a short, others for a long time. Which in the oceans are the electric fields sharks and rays detect? How do the sense organs, the ampullae of Lorenzini, operate? How do we biophysically explain their astounding nV/cm electrical sensitivity? How are the sensory data processed centrally? How do the animals interpret the sensory signals they receive from the oceans' electric and magnetic fields? How do they proceed to arrive at the source of a prey field? Can they electrically detect objects that do not produce but only distort electric fields? What is the role of their electric and magnetic sense in navigating ocean waters? Lingering uncertainty, controversy, and misconceptions have since in part been resolved by new observations and sheer reflection. Yet, much more remains to be discovered.

Regarding the localization of underwater objects, after years of reflection, two long-belated ideas have emerged in my mind. Given the impossibility for a shark to sense from a distance where the source of a prey's electric multipole field is, I have conceived of and mathematically formulated a guided-approach algorithm leading sharks along well-defined paths to their targets without having to know where the target is (Kalmijn 1988a). Moreover, I have inferred that sharks not only detect objects that produce electric fields, but also objects that merely distort existing ambient electric fields, hence producing image fields, and that they may do so based on the very same approach algorithm. To prove these enticing ideas to be realistically feasible, I have successfully tested sharks on their ability to detect—in the presence of an ambient electric field—electrically insulating objects hidden under the loose sand on the bottom of a large fiberglass observation tank, and I have computer modeled and hardware implemented a shark-inspired autonomous robotic underwater device capable of approaching and attacking objects based upon the inferred electrically guided approach algorithm.

Physical aspects

To apprehend the sharks' electric and magnetic sensory abilities, a solid grasp is needed, not only of the functioning of their elaborate electroreceptor system and the central processing of the sensory data, but also of the mathematical physics defining the biologically significant features of the oceans' electric and magnetic fields. In orienting in the earth's magnetic field, sharks predated Einstein's line of thinking. When studying motionally induced electric fields, we used to call on Faraday's empirical law. However, the special theory of relativity tells us that moving in a magnetic field in the earth frame of reference means for a shark, in whose frame of reference the sense organs are at rest, *"nothing else but"* receiving a uniform electric field. Einstein even might have referred to sharks instead of trains in his 1905 paper, if he had known of their motional electric magnetic sense. Hence, *to a shark, the electric fields of ocean currents and the electric fields resulting from its own motional activity appear both as imposed uniform electric fields that differ only in their dependence on the animal's exploratory movements.* This teaches us that, to fully appreciate a shark's perception of the subsea world, we must interpret the electrical signals it receives in its proper frame of reference.

The ampullary system detects the electric fields differentially by sensing the potential differences between the large number of electroreceptive skin pores, on the order of a thousand, which in sharks are confined to the head region, but in rays also spread over the wing-like pectoral fins and the body down to the base of the tail. Since the electric fields of ocean streams and the field of the shark's own motions in the earth's magnetic field all appear to the animal as essentially uniform, sharks may use the whole complement of three-dimensionally oriented ampullae of Lorenzini to establish just one single defining uniform-field vector at a central point in the head. Also, in detecting prey from a distance, the approach algorithm guiding sharks to their target operates on the locally nearly uniform bioelectric prey fields of their targets, responding to the change in the direction of the uniform-field vector as the animal proceeds. However, upon nearing the target, the more and more noticeable non-uniformity of the potential distribution over the electroreceptive skin area becomes significant as well, which sharks and rays, in the *proximity mode,* may use for the final localization of the target.

To interpret the electric fields sharks receive in ocean streams from both the flow of water and from their own motions in the earth's magnetic field, we must assume that the animals discriminate between the components of the signals acting along their three mutually orthogonal body axes for us to understand their processing. Hence, we will express the electric-field signals the central nervous system receives from the ampullae in their components along the aft-fore or caudal-rostral, the down-up or ventro-dorsal, and the left-right or transverse body axes and tacitly assume the central nervous system also to process the incoming electric-field data three-dimensionally, regardless of how it may do so neurophysiologically. Obviously, the individual ampullae, being in general not strictly aligned with the body axes, carry variously oriented signals. However, not just the individual electroreceptors, but the sensory system as a whole, gathers and subsequently processes the electrical information in the format needed for the desired behavioral responses to be executed three-dimensionally.

But why does the present article introduce the physics of electric and magnetic orientation from the earth-bound observer's point of view, if it is the animal that must do the orienting? The answer is of a psychological nature: because that is how we earthlings experience the universe, as if we still are at the center of it. Rest assured, though, from a physics standpoint, there is nothing wrong with this approach, and I will follow it to render the article as accessible to the reader as possible. By correctly reasoning from our point of view, we will without great effort arrive at the physically adequate stimulus the animal receives in its own frame of reference. It will be not difficult for us to transform the results to the shark's frame. Lorentz and Einstein told us how. The fortunate result is that, in both the earth frame and the shark frame, the physically adequate stimulus is $-\rho\mathbf{J}$ integrated across the body, with $\mathbf{J}$ the density of the motionally induced electric current through the animal's body tissues and skin, bypassing the ampullary sense organs. Hence, it will indeed be quite feasible to view the underwater world electrically the same as sharks and rays do. I have learned to do so over the years, and I encourage the reader also to make a serious attempt to pretend to be a shark.

The present discourse focuses on the physical theory of electric and magnetic orientation. The formal, comprehensive version applies to all kinds of animals, from sharks to birds, dwelling in all kinds of media, from seawater to thin air. Yet, for the magnetic orientation of sharks in ocean waters, I also have offered a less exacting, more intuitive version, which may suffice for the physics-less-inquisitive reader. However, the pared-down version must be applied with great care. Even for marine sharks it is only approximately valid, but to freshwater fish and airborne birds it does not apply at all. In studying freshwater fish or migrating birds, scientists unaware of this caveat may sow more confusion than clarity. The shortcoming of the simpler version is that it neglects the charge accumulations on opposite skin surfaces of the animal, which in the specific instance of marine sharks are rather small, though never nil, making it look as if the $\mathbf{v} \times \mathbf{B}$ is the physically adequate stimulus, instead of $-\rho\mathbf{J}$, though it is not. True, the $\mathbf{v} \times \mathbf{B}$ passes on the orientational information, but it is the $-\rho\mathbf{J}$ resulting from the $\mathbf{v} \times \mathbf{B}$ that the animal detects, an often-misunderstood issue of fundamental importance. Moreover, the simpler version only considers the view of an observer in the earth frame, whereas the formal theory emphasizes the view of the shark in its own frame of reference. Sharks would have no notion of what the simpler version is about. In their own frames, they are not moving, but the whole world is. Hence, to grasp the orientational challenges a shark faces, its own frame is the proper one.

In my (Kalmijn 1974) handbook chapter, I stated "Motional-electric fields are induced whenever water is moving, or a fish is swimming through the earth's magnetic field." I could have added "You know what I mean?" But what does moving through a magnetic field really mean? A magnetic field is a condition, not a thing, not a wire cage of field lines. Two observers moving with respect to each other measure the same condition magnetically differently. True, those differences are trivial at the low speeds of motion we deal with. However, when the two observers measure the consequences of an object moving in the presence of a magnetic field in their respective frames of reference, the results are very different, as the velocity $\mathbf{v}$, hence $\mathbf{v} \times \mathbf{B}$, of the moving object for them is very different. Thus, the motionally induced electric field $\mathbf{v} \times \mathbf{B}$ can have any value depending on the choice of reference frame in which the observer is at rest. In particular, for a shark at rest in its own frame, its $\mathbf{v} \times \mathbf{B}$ is zero, as was the crux of the matter in the last three lines of the previous paragraph. In short, moving in a magnetic field or through a magnetic field means moving with respect to a frame of reference in which there is a magnetic field. Moving with respect to a magnetic field is nonsense, though one may speak of moving with respect to the source of a magnetic field.

Before we proceed, the meaning of the term "motional electric field" must be agreed upon. It is often used generically for any electric field resulting from relative motion. In the present discourse, it may refer to (1) the *motionally induced electric field* $\mathbf{v} \times \mathbf{B}$ acting on the charges of the ionic particles moving in a magnetic field, (2) the *motional-electric field* **E** *in the strict sense* due to the charges accumulated by the electric current resulting from the motionally induced electric field, or (3) the *ohmic field* $-\rho\mathbf{J}$ of the resistive drag force field experienced by the moving charged ionic particles. The three are closely related, yet fundamentally different. Actually, the $-\rho\mathbf{J}$ field is, strictly speaking, not an electric field, as it does not act on the charges, but on the particles carrying the charges. The qualifier *in the strict sense* is commonly omitted. Usually, it is clear from the context what is meant by *motional electric*. Nonetheless, adding the symbols $\mathbf{v} \times \mathbf{B}$, **E**, and $-\rho\mathbf{J}$ to the terms may remove the ambiguity. The reader is advised to check what authors of physics texts and journal articles mean by the terms they use, as little consensus consists, sometimes even about the symbols representing them.

Furthermore, unnecessary disagreement has resulted from the use of imprecisely defined terms denoting the physical nature of the motional electric fields the animals detect. What do we mean in the present article by stating: that *in sensing their drift and compass heading, sharks detect the* $-\rho\boldsymbol{J}$ *integrated across the body*? To settle the issue, let us agree: (1) *the biologically adequate stimuli* are the oceans' electric and magnetic fields sharks receive biologically meaningful information from, regardless of how the animals detect the fields, (2) *the physiologically adequate stimuli* are the electric fields exciting the ampullary sense organs in their natural amplitude and frequency range, whether or not they are biologically adequate, including man-made square-wave or sinusoidal test stimuli, and (3) *the physically adequate* or *effective stimuli* are the fields offering a quantitative measure of the excitation of the sense organs. The cardinal question is: which of those is the physically adequate stimulus: the *motionally induced electric field* $\mathbf{v} \times \mathbf{B}$, the *motional electric field in the strict sense* **E**, or perhaps the formally non-electric *ohmic field* $-\rho\mathbf{J}$, integrated across the body?

As you will learn, perhaps to your surprise, in the detection of ocean streams and magnetic compass headings, the physically adequate stimulus is not defined by the motionally induced electric field $\mathbf{v} \times \mathbf{B}$, nor by the motional electric field **E** of the charge accumulation on opposite sides of the stream or the animal, but is defined by the $-\rho\mathbf{J}$ field of the ohmic resistive forces resulting from the induced electric current of density **J** in the medium of resistivity ρ. Why? Because the physically adequate stimulus is given by the line integral of $\mathbf{v} \times \mathbf{B}$ *plus* **E**, the sum of which equals $-\rho\mathbf{J}$, whatever the motionally induced electric field $\mathbf{v} \times \mathbf{B}$ and the electric field of the charge accumulations **E** may be, not only in marine but also in freshwater and airborne animals. Hence, of the three only the $-\rho\mathbf{J}$ correctly measures the excitation under all circumstances. This is one of the main tenets of the first section of the present article. One proviso shall be mentioned: the additional excitation caused by the charges accumulating on the skin of a shark drifting in an ocean stream is not considered part of the physically adequate stimulus, but part of the sensory system enhancing the physically adequate stimulus before its detection by the sense organs, as will be made clear in a later section.

Since the biologically oriented reader may not have received the training needed to grasp the intricacies of the physics-based electric and magnetic orientation theory and even physicists may need some refreshing, I have added a few physics and mathematics notes, embedded in the text and further discussed in scholia. The selection of these appendices was made on the basis of the issues I myself initially had difficulty with. I hope, I got them right now. The physical and mathematical stumbling blocks generally encountered are obvious from the several sadly erroneous literature summaries of my electric and magnetic orientation theory. In particular, how the forces exerted by electric fields relate to the gradients of the fields. How do potentials relate to potential gradients and voltages to voltage gradients? What are the differential and integral properties of the fields? What are the proper names and mathematical symbols of the fields? To hearten the reader, I readily admit, it did not come easily to me either. I have worked on the theory for over half a century, gradually developing it since my discovery of the astounding electrical sensitivity of sharks and rays. Even now, I have some lingering doubts, which I hope the next generation of scientists will remove. Here, I take it as my holy task to document the theory in its most complete form after all those years I have worked on it.

1.3 Behavioral Evidence of Electric and Magnetic Orientation in Sharks and Rays

For references, see IV-a. Author's annotated selected publications.

Electric and magnetic orientational capabilities and sensitivity levels
From the behavioral experiments I conducted in my marine research facilities and at sea over the years (Kalmijn 1971, 1982), it has become evident that sharks and rays meaningfully respond to naturally occurring DC and low-frequency AC *electric* fields. In the wild, the shark *Mustelus canis* initiated its night-time feeding attacks at electrically simulated prey at distances from the source at which the DC field was as weak as 1–2 nV/cm. In my land-based animal habitat, the shark *Scyliorhinus canicula* performed clear feeding responses to dipole fields at frequencies from DC up to 4 Hz. Furthermore, in protracted series of conditioning experiments, the stingray *Urolophus halleri* proved not only to detect the presence of the oceans' virtually uniform ambient DC electric fields, but also to rely on their direction and polarity (Fig. 1.1). *The statistically highly significant results manifestly established the stingray's ability to orient in uniform electric fields of 5 nV/cm.* In a wide, shallow bay, the shark *Negaprion brevirostris* and the ray *Dasyatis pastinaca* confirmed the results in that they changed their rectilinear paths of travel by 60 degrees when the ambient electric field was instrumentally turned 60 degrees and reassumed their original course of swimming when the field was relaxed back to normal. Moreover, acoustic trackings of the blue shark *Prionace glauca* in the Atlantic were never observed to oppose the drift of water (Carey et al. 1990).

The evoked behavioral responses of the shark *Triakis semifasciata* to magnetic fields became apparent from its fierce attacks directed at a small permanent bar magnet held under the bottom of its fiberglass seawater basin and from its frantic avoidance reactions to a DC *magnetic* field produced by an 8″ in diameter induction

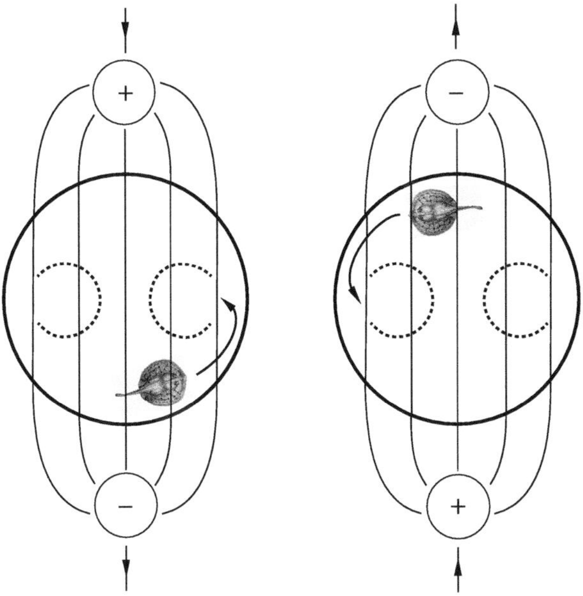

Fig. 1.1 Stingray *Urolophus halleri* seeks food in enclosure to the left of ambient *electric* field under both normal and reversed electric field conditions

Fig. 1.2 Stingray *Urolophus halleri* seeks food in enclosure in the East of ambient *magnetic* field under both normal and reversed magnetic field conditions

coil held outside the cylindrical wall of the shallow tank, locally distorting the earth's magnetic field just inside the tank for about 25%. *These observations suggested that sharks moving in a magnetic field induce electric fields, which, if mere local, they may mistake for the bioelectric fields of prey or otherwise avoid in fright.* Yet, to prove a genuine magnetic sense, really meaningful orientational responses in the earth's magnetic fields or faithful simulations thereof were required. Hence, I conducted long series of conditioning experiments, in the same fashion as the ambient electric field experiments but this time with an ambient magnetic field, in which the stingray *Urolophus halleri* proved not only to detect the earth-strength ambient magnetic field controlled by large Helmholtz coils, but also to be able to use the direction and polarity of the field to decide on which side of their seawater basin, either in the east or the west, to secure food (Fig. 1.2).

The highly significant results unequivocally established the animals' ability to orient in the earth's magnetic field, not only in the inclined magnetic fields of temperate waters, but also in the purely horizontal field along the magnetic equator.

Similarity and difference between electric and magnetic orientational cues
Before discussing the electric and magnetic orientation of sharks and rays in greater depth, it must be made clear that the animals' orientational abilities were theoretically anticipated from their phenomenal electrically sensitivity and the physical principle of motional electric induction several years before the behavioral experiments corroborating their biological implementation. Orientation in the motion-electric fields of ocean streams and the animals' detection of their magnetic compass headings may seem very different endeavors. Nevertheless, on close inspection, in the sharks' frames of reference, the two appear equally feasible, as both are received as uniform electric fields, merely differing in their dependence on the animals' exploratory movements. If the electrical function is in operation, then it is extremely unlikely that the motional-electric magnetic function is not. Indeed, it is endlessly more demanding, to argue that the electric field detection is enacted, but the magnetic field detection is suppressed, or *vice versa.* Even so, there are confused scientists who prefer the sharks' motional electric sensitivity to the earth's magnetic field biologically to be regarded as a nuisance rather than an asset, as it would interfere with the detection of prey.

Electrophysiological versus behavioral sensitivity
From my graduate-student investigations, it is evident that the ampullae of Lorenzini are the sense organs mediating the described behavioral responses to electric and magnetic fields. Recently, I have also gained more insight into the biophysical principle of operation of the sense organs, as will be briefly touched upon later on in the present article. Admittedly, the electrical sensitivity judged from afferent nerve responses, even when taking the large number of ampullary sense organs into account, has not yet met the behavioral sensitivity. But the remaining difference does not seem to negate either one, as it has not yet adequately explained how the animal processes the nerve action potentials and central-nervous signals to enhance sensitivity. It certainly is not a shortcoming of the animals, but rather of our interpretation of the electrophysiological data. The nerve recordings are in general made in response to artificial stimuli, sinewaves, pulses, or steps. *The responses to naturally occurring signals, though, may well be processed based on the stimuli expected to be received,* thus rendering the application of more powerful processing algorithms quite feasible.

Comparison with the ampullary electric sense of freshwater catfishes
As an interesting note, based on what I had learned from the sharks and rays, I wondered whether catfishes and weakly electric fishes, known also to have low-frequency ampullary electroreceptors—the latter in addition to the higher-frequency tuberous electroreceptors—might as well detect the DC bioelectric fields of prey and orient in the DC ambient electric fields of fresh waters. In long series of behavioral experiments, I have proven both the bioelectric prey detection and the

ambient-electric field orientation in these freshwater electrosensitive fishes indeed to hold as well. The main difference being that catfishes and weakly electric fishes are about a hundred times less sensitive electrically. Yet, the DC bioelectric and ambient electric fields they detect are about a hundred times stronger. Moreover, the freshwater ambient electric fields are not so much of a motional electric but of an electrochemical origin. A drawback for the freshwater fishes is that they are not sensitive enough to orient motion-electrically in the earth's magnetic field. As for the receptor cells, in the ampullae of Lorenzini the apical membrane with its lone kinocilium drives the response to the stimulus, whereas in catfishes and weakly electric fishes it is the basal membrane that drives the response.

To get to the heart of the matter

The first section of the ensuing discourse introduces the electric and magnetic fields informing sharks of their drift in ocean streams and their magnetic compass headings, tacitly assuming that they detect the fields as presented to them. The physical theory *per se* is either correct or wrong, but not a matter of taste. The second section discusses the exploratory movements sharks may make to detect and discriminate between the fields of ocean streams and those they themselves motion-electrically induce. Again, the logic will be either correct or wrong. However, whether the animals follow the proposed procedures is merely a matter of inference based on logical reasoning. Hence, I do not claim sharks to behave as envisioned by me, although they may as well. My reasoning is only meant to show the physical and biological feasibility of sharks electrically sensing their drift and magnetic compass heading side by side. The third section should not have been needed at all. Yet, according to my strict, old-fashioned European education, I consider it my scientific obligation to tell the reader not only why what I believe is right but also to explain why what I do not believe is wrong, not wrong because I say so but wrong according to well-established facts and knowledge. It makes me sad to learn about the toxic, wide off the mark darts my pioneering investigations have received. I hope those who threw them will come to their senses, to save science from further embarrassment.

References

Carey FG, Scharold JV, Kalmijn AJ (1990) Movements of blue sharks (*Prionace glauca*) in depth and course. Mar Biol 106:329–342

Einstein A (1905) Zur Elektrodynamik bewegter Körper. Annalen der Physik 17:891–921. See also: On the electrodynamics of moving bodies. In: The Principle of Relativity. Translated by W. Perrett and G. B. Jeffery, 1952, Dover edition, pp. 35-65

Faraday M (1832) The Bakerian Lecture: experimental researches in electricity. Second Series. Phil Trans R Soc Lond 122:163–194. See also: 1965 Dover edition of Experimental Researches in Electricity. Vol. I, pp. 42-75

Galileo (1632) Dialogica translated by S. Drake 1967, Dialogue concerning the two chief world systems, pp 186–187

Heaviside O (1893) 1899, 1912 Electromagnetic Theory. In three volumes. 1971, Chelsea Publishing Company, New York. Beyond Maxwell

Kalmijn AJ (1971) The electric sense of sharks and rays. J Exp Biol 55:371–383

Kalmijn AJ (1974) The detection of electric fields from inanimate and animate sources other than electric organs. In: Fessard A (ed) Handbook of sensory physiology, vol III/3. Springer, Berlin, pp 147–200
Kalmijn AJ (1982) Electric and magnetic field detection in elasmobranch fishes. Science 218:916–918
Kalmijn AJ (1988a) Detection of weak electric fields. In: Atema J, Fay RR, Popper AN, Tavolga WN (eds), Sensory Biology of Aquatic Animals, pp. 151–186. Springer-Verlag, New York
Kalmijn AJ (1988b) Hydrodynamic and acoustic field detection. In: Atema J, Fay RR, Popper AN, Tavolga WN (eds), pp. 83–130. Springer-Verlag, New York
Lorentz HA (1892) La Théorie électromagnetique de Maxwell et son application aux corps mouvants. Leide, E. G. Brill
Lorenzini S (1678) Osservazioni intorno alle Torpedini. Firenze, pp. 7–11, Plate I–II, pp. 22–24, 26–28
Maxwell JC (1891) A treatise on electricity & magnetism. Unabridged Third edition. 1954 Dover edition in two volumes. Mathematical foundation

Chapter 2
Passive and Active Modes of Electro-orientation

The idea of sharks orienting in the oceans' ambient electric fields and—motional electrically—detecting the earth's magnetic field did not occur to me to explain how the animals navigate the vast ocean waters. On the contrary, it was their extraordinary electrical sensitivity for which I sought a biological function. Historically, their electrically guided predatory attacks incited me to conclude that sharks and the kindred rays have a genuine electric sense. Moreover, when I found them to exhibit clear behavioral responses to DC and low-frequency AC electric fields of merely a few nanovolt/cm, it became evident that sharks day and night are exposed to electric fields well within their sensitivity range, not only when passively drifting in ocean streams, but also by actively moving in the earth's magnetic field. Thus, the question was not whether sharks detect the oceans' electric and magnetic fields, from which they cannot escape, but whether they indeed use them to orient in open ocean streams and to sense their magnetic compass headings, orientational abilities that would serve them well in traveling ocean waters and in moving about in their preferred territory without drifting off.

One may deem the task of orienting in the vast electric fields of ocean streams or in the global earth's magnetic field not implausible, since sharks have ample opportunity to average and interpret the electrical signals they receive over time and distance. That may well be true. However, the same information is needed instantly, when the animals, motivated by smell, are frantically searching for food and all of a sudden execute their fast, electrically guided approaches to prey items in the presence of the ambient electric and magnetic field. These predatory responses require sharks all the time to monitor the signals they receive from interaction with the ambient electric and earth's magnetic fields and to take them into account when executing their electrically guided feeding attacks. I consider this the most amazing feat of the sharks' electric sense. Yet, similar situations present themselves on a regular basis with other sensory systems, as we all know, for instance, when jumping across a creek taking gravity into account, when walking in the woods looking out for trees and mountain lions, or as a pilot flying across the ocean correcting for

A. J. Kalmijn, *Theory of Electric and Magnetic Orientation in Sharks and Rays Revisited*, https://doi.org/10.1007/978-3-031-58445-9_2

sideways drift. Clearly, sense organs seldom operate in isolation. Hence, the detection of the ambient electric as well as the earth's magnetic field shall be very fast, faster than the prey can detect and escape from the shark.

To expose the physical principles of electric and magnetic orientation, we will start from the electrical signals a shark receives when passively drifting in an open-ocean stream and when actively traveling in the earth's magnetic field *at constant, rectilinear velocities*. True, the stimuli the shark thus receives are essentially of a DC nature, whereas its electroreceptors are not DC sensitive. Yet, in the present section we will learn that the DC electrical stimuli the animals receive from interaction with the DC ambient electric and magnetic fields contain all the information needed for the animals to orient in the oceans' electric and magnetic fields, and for now ignore the AC nature of the sense organs, assuming that the animal anyhow knows how to detect DC fields. In the second section, we will detail the exploratory movements I envision the animal may execute to transfer the DC stimuli into its AC frequency range. As we will learn, the animal can *detect* the DC fields it *receives* in ocean streams by turning left or right and *detect* the DC electric fields it *receives* from interaction with the earth's magnetic field by rectilinearly accelerating or decelerating. Also, the sharks' characteristic meandering mode of swimming and the periodic motions between body parts, such as the tail and fin movements, will be considered. In the third section, we will inquire into alternative opinions and make an effort to separate the wheat from the chaff so as to pave the road to progress and to stop the endless bickering that may leave the newcomer in doubt of who and what to believe.

An issue of fundamental importance, but often shunned, shall be elucidated before we enter into a detailed discussion: in which frame of reference does a shark orient in the electric fields of ocean streams and in the ubiquitous earth's magnetic field? Physically speaking, animal and man execute the process of orienting in their own, proper frames of reference, that is, they detect and process the sensory signals they receive in the frames of reference in which they are at rest. However, we customarily transform the sensory information we receive in our proper frames to the earth frame and are hardly aware of it. When our eyes see the landscape move by, we merely *infer* that we are the ones moving. Indeed, we habitually transform most, though not all, sensory data from the proper to the environmental frame of reference without realizing how great an achievement it is. The world, as we see it in three dimensions, is merely the interpretation of two tiny two-dimensional images on our retinas. The world, as sharks electrically sense it in three dimensions, is merely the interpretation of the potential distribution over their closed electroreceptive skin area. How they perceive the electrical and magnetic world is a valid question, but we can only guess the answer.

For sharks, this is not a matter of semantics. Even if they would be rational, sharks cannot know more than they perceive by use of their senses. By detecting magnetic fields motion-electrically, magnetic fields would for sharks be an abstract inference. Hence, in their frames of reference, the environment moving in a global magnetic field is not a sensible concept. They themselves moving in their own rest frame is a logical impossibility. Nevertheless, we accredit sharks with a magnetic sense, since they are capable of orienting in—what we call—the earth's magnetic field, regardless of how they do so. To resolve the dilemma, we must understand that, in fact, sharks in their own frame of reference are subject to uniform electric fields when the world at large moves with respect to them, either passively or as a result of their own activity, and apparently have learned to use these electric fields to sense—what we call—their drift and compass heading. Yet, we will start our discussion from the earth-bound observers' point of view, not the sharks', in order to approach the issue based on familiar scientific knowledge and to prepare ourselves for the sharks' point of view. However legitimate our self-centered notion of a shark moving through the water may be, do not forget that ultimately it is the shark who must do the sensing and orienting. I study sharks, not scientists.

Animal navigation biologists are intrigued by the detection of the earth's magnetic field. Do animals detect the earth's magnetic field as such in its full strength, or do they detect only the local nonuniformities in the direction and strength of the earth's magnetic field, or perhaps both? As for sharks, they for sure receive electrical information about the earth's magnetic field in its full strength owing to the motionally induced electric fields they generate when moving relative to the water. It is extremely unlikely, though, that sharks motion-electrically detect the local magnetic nonuniformities, as those would demand considerably greater electrical sensitivity than we can reasonably expect. Of course, they may well have a second magnetic detection system operating on other physical principles to detect the nonuniformities, for which there is at the moment no convincing evidence. Instead, sharks may rely on the global pattern of ambient-electric fields associated with the worldwide ocean circulations, which may serve a similar purpose for navigation as the magnetic anomalies.

Below, we will first describe the physical situation in terms of *fields and forces at single points*, when, after a brief period of acceleration, the driving electrical and the frictional drag forces acting on the ionic particles have reached their dynamic equilibrium and the electric currents have become stationary. This local description focuses on the motions of the individual ions and, ultimately, on the forces acting on the gating charges of the voltage-sensitive ion channels. Subsequently, we will discuss the situation in terms of *voltages and potential differences between different points*. This spatially extended description lends itself better to assessing the effective strengths of the fields, in particular the strength of the electrical stimulus presented to the animals. This dual approach has been chosen to satisfy the conceptually oriented as well as the mathematically inclined readers. Much confusion has resulted from not clearly distinguishing between the two approaches, point by point or from point to point, as the author has experienced himself. The present discourse emphasizes first principles. It is kept as simple as possible—but not simpler—by idealizing the physical situations without loss of validity. Realistic assumptions are made so as to focus on the bare fundamentals. Yet, the constraints will later on be relaxed to conform to the real, incalculably intricate subsea world.

2.1 Passive Mode of Electro-orientation: Shark Detecting Drift in Ocean Stream

2.1.1 Physical Principles of Passive Electro-orientation: Forces on Ionic Charge Carriers[1]

2.1.1.1 Shark Drifting in Open-Ocean Stream (Fig. 2.1)

Imagine a shark drifting in a shallow, wind-driven ocean stream flowing with a velocity $\mathbf{v}$ relative to still, deeper water. Might the animal derive the upstream direction and the speed of the stream from the electrical stimuli it receives from the motionally induced electric field $\mathbf{v} \times \mathbf{B}_v$ that the flowing water and the drifting shark give rise to by interaction with the vertical component $\mathbf{B}_v$ of the earth's magnetic field? Let us first ask a scientist at rest in his familiar earth frame of reference, who observes the stream, the shark, and the animal's sense organs moving in a magnetic field, and then the shark in its own, proper frame of reference, for which the stream, the animal itself, and its sense organs are at rest. We will pretend to be in the southern hemisphere, where $\mathbf{B}_v$ points up and, hence, the vector product $\mathbf{v} \times \mathbf{B}_v$ is directed horizontally, normal to the upstream direction of the stream, as shown in the vector diagram of Fig. 2.1. To free the discussion from non-essential intricacies, we will

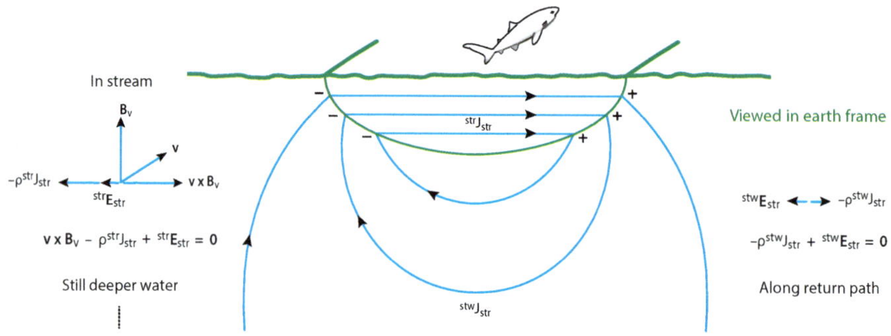

Fig. 2.1 Shark drifting in wind-driven, open-ocean stream. Stream flowing with velocity $\mathbf{v}$ creates, by interaction with vertical component of the earth's magnetic field $\mathbf{B}_v$, horizontally directed *motionally induced electric field* $\mathbf{v} \times \mathbf{B}_v$. Resulting *induced electric current* of density $^{str}\mathbf{J}_{str}$ accumulates – and + space charges at lateral boundaries of stream, creating *motional-electric field* $^{str}\mathbf{E}_{str}$. Left: $^{str}\mathbf{J}_{str}$ caused by $\mathbf{v} \times \mathbf{B}_v$ is impeded by *ohmic drag force* $-\rho^{str}\mathbf{J}_{str}$, with ρ resistivity of seawater, and by $^{str}\mathbf{E}_{str}$ of space charges at boundaries of the stream, hence, in the stream, $\mathbf{v} \times \mathbf{B}_v - \rho^{str}\mathbf{J}_{str} + {}^{str}\mathbf{E}_{str} = 0$. Right: electric field of + and – space charges furthermore drives electric current of density $^{stw}\mathbf{J}_{str}$ along return paths through still, deeper water, along which $-\rho^{stw}\mathbf{J}_{str} + {}^{stw}\mathbf{E}_{str} = 0$. Lengths of $\mathbf{v} \times \mathbf{B}_v$, $-\rho\mathbf{J}_{str}$, $-\mathbf{J}_{stw}$, $\mathbf{E}_{str}$, and $\mathbf{E}_{stw}$ arrows indicate stream-wide averages of *forces per unit charge on positive ions*. To obtain forces on individual ions, they must be multiplied by signed charges of ions

[1] The author recommends consulting Appendix 2 now, or to come back to it at a later time, to ensure the reader understands what the mathematical symbols stand for and the formulas physically express.

assume the velocity of the stream **v**, the magnetic field $\mathbf{B}_v$, and the resistivity of the seawater ρ to be uniform and independent of time, and furthermore the stream to have a semi-elliptical cross-section and not to vary along its length. Interaction with the horizontal component $\mathbf{B}_h$ of the earth's magnetic field will be discussed later. How a shark detects the DC fields with its AC receptors will be explained in Sect. 3.2. Here, we will focus on the electrical signals the animal receives and the information they present.

2.1.1.2 In the Undisturbed Stream, Well Away from the Shark (Fig. 2.1)

The horizontally directed *motionally induced electric field* $\mathbf{v} \times \mathbf{B}_v$ exerts a force on the ionic charge carriers in the stream, causing an *induced electric current* of density ${}^{str}\mathbf{J}_{str}$ to flow across the stream, with the positive ions traveling in the direction of $\mathbf{v} \times \mathbf{B}_v$ and the negative ions in the opposite direction. At the lateral boundaries of the stream, where the velocity of the water **v** and, hence, $\mathbf{v} \times \mathbf{B}_v$ changes steeply, the ions entering the flowing water accelerate, whereas the ions leaving the flowing water decelerate, thus accumulating – and + space charges on opposite sides of the stream. Hence, in the stream, the electric current is impeded by two forces: the *ohmic drag force* $-\rho^{str}\mathbf{J}_{str}$, with ρ the resistivity of the seawater, and the *motional electric field* ${}^{str}\mathbf{E}_{str}$ of the charges at the lateral boundaries of the stream, whereas in the still water adjacent to and under the stream the field ${}^{stw}\mathbf{E}_{str}$ of the space charges causes an electric current of density ${}^{stw}\mathbf{J}_{str}$, which returns the electric current induced in the stream along paths through deeper water and the ocean bottom, completing the electrical circuit. The space-charge-increasing current of density ${}^{str}\mathbf{J}_{str}$ across the stream and the space-charge-diminishing current of density ${}^{stw}\mathbf{J}_{str}$ along the still-water return paths adjust the – and + charges at the lateral boundaries of the stream so as to render the electric current continuous around the closed circuit across the stream and back along the return path.

Note that the minus sign in $-\rho\mathbf{J}_{str}$ is part of the symbol, indicating that the direction of the resistive force $-\rho\mathbf{J}_{str}$ is opposite to that of the current density $\mathbf{J}_{str}$, with ρ the resistivity of the medium at the point in the field that **J** pertains to. Note also that ${}^{str}\mathbf{E}_{str}$ and ${}^{stw}\mathbf{E}_{str}$ apply to different regions of the same field, to wit, the field of the charge accumulations at the lateral boundaries of the stream. The two regions differ in that the stream is in motion and is characterized by the presence of $\mathbf{v} \times \mathbf{B}_v$, whereas the return path remains still and is characterized by the absence of $\mathbf{v} \times \mathbf{B}_v$.

In the stream, the motionally induced electric force $\mathbf{v} \times \mathbf{B}_v$, the ohmic drag force $-\rho^{str}\mathbf{J}_{str}$, and the motional electric force ${}^{str}\mathbf{E}_{str}$ due to the charge accumulations at the lateral boundaries of the stream are, in the stationary state, in dynamic equilibrium, $\mathbf{v} \times \mathbf{B}_v - \rho^{str}\mathbf{J}_{str} + {}^{str}\mathbf{E}_{str} = 0$, whereas along the still-water return paths, where **v** and, hence, $\mathbf{v} \times \mathbf{B}_v$, are zero, solely the two, $-\rho^{stw}\mathbf{J}_{str}$ and ${}^{stw}\mathbf{E}_{str}$, are in dynamic equilibrium, $-\rho^{stw}\mathbf{J}_{str} + {}^{stw}\mathbf{E}_{str} = 0$. In dynamic equilibrium means that the positive and negative ions have reached their limiting velocities after a brief, transient acceleration, during which the ion-velocity dependent ohmic $-\rho^{str}\mathbf{J}_{str}$ and the motional electric ${}^{str}\mathbf{E}_{str}$ of the electrostatic charge accumulations at the boundaries of the stream have grown so strong as to balance the motionally induced electric

$\mathbf{v} \times \mathbf{B}_v$. The amounts of charges building up at opposite boundaries of the stream depend on how readily the ionic charges are carried off along the return path through deeper water and the ocean bottom, hence, on the resistance of the return paths. After the stationary state has been reached, no further changes take place at constant $\mathbf{v}$, $\mathbf{B}_v$, and ρ. In our lengthwise uniform, in cross-section semi-elliptic model stream, chosen for its unique motional electric simplicity, not only the $\mathbf{v} \times \mathbf{B}_v$, but also the resulting $-\rho^{str}\mathbf{J}_{str}$ and ${}^{str}\mathbf{E}_{str}$ are inherently uniform at steady and uniform $\mathbf{v}$ and $\mathbf{B}_v$.

2.1.1.3 Inside and in the Vicinity of the Drifting Shark (Fig. 2.2)

After introducing the shark into the ocean stream, the motionally induced electric field $\mathbf{v} \times \mathbf{B}_v$, as well as the motional electric field ${}^{anl}\mathbf{E}_{str}$ in the drifting shark and the motional electric field ${}^{str}\mathbf{E}_{str}$ in the vicinity of the animal resulting from the – and +

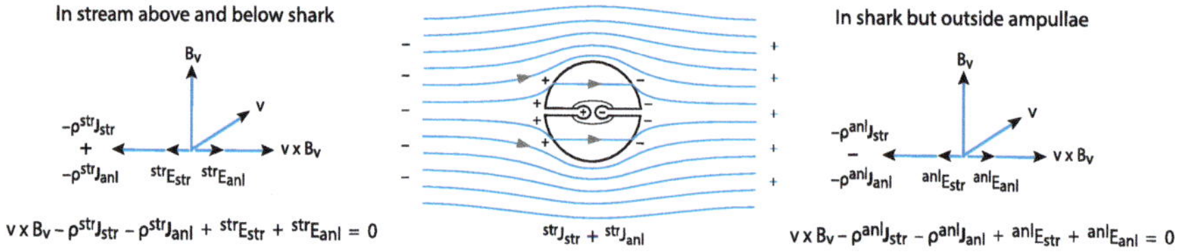

Fig. 2.2 Dynamic equilibrium of forces acting on positive ionic charges in stream and drifting shark. Middle: cross-section of shark with differential pair of ampullary sense organs; ampullae proper enclosed in single or paired ampullary capsule; – + space charges at lateral boundaries of stream; + – surface charges on opposite skin surfaces of animal; + – surface charges on epithelia lining ampullary canals and blind ampullae proper; electric current of density ${}^{str}\mathbf{J}_{str}+{}^{str}\mathbf{J}_{anl}$ flowing outside past animal, deflected by distributed dipole field ${}^{str}\mathbf{E}_{anl}$ of charge accumulations on skin, with ${}^{str}\mathbf{J}_{anl}$ *aiding* ${}^{str}\mathbf{J}_{str}$; electric current of density ${}^{anl}\mathbf{J}_{str}+{}^{anl}\mathbf{J}_{anl}$ flowing through skin and body tissues inside animal, with ${}^{anl}\mathbf{J}_{anl}$ *opposing* ${}^{anl}\mathbf{J}_{str}$. Left: vector diagram in stream above and below shark: $-\rho^{str}\mathbf{J}_{str}$ and $-\rho^{str}\mathbf{J}_{anl}$ add. Right: vector diagram in shark but outside ampullary sense organs: $-\rho^{anl}\mathbf{J}_{str}$ and $-\rho^{anl}\mathbf{J}_{anl}$ subtract. Not to scale, lengths of arrows indicate average *forces per unit charge on positive ions*

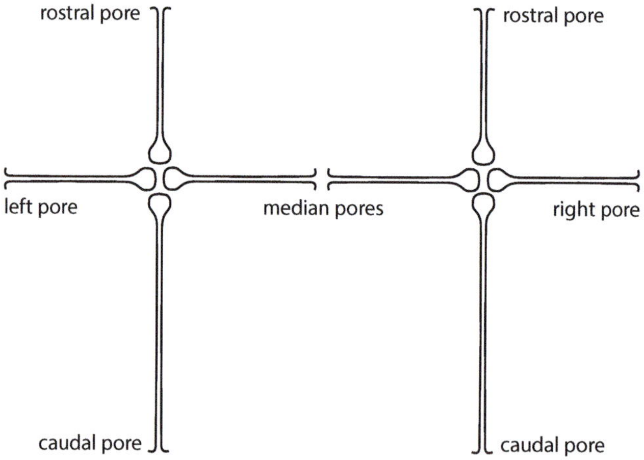

charges at the lateral boundaries of the stream remain the unaltered. Yet, this does not mean that the fields in the two regions do not change. It only means that the contributions of $\mathbf{E}_{str}$ to these fields remain the same. However, since an oceanic shark is on the average about a hundred times more resistive than the seawater it displaces, ≈ 2 kohm cm versus ≈ 20 ohm cm, the motionally induced electric current of density ${}^{str}\mathbf{J}_{str}$ flowing across the stream causes in addition + and – charges to accumulate on opposite skin surfaces of the shark, impelling most of the motionally induced electric current in the stream to *flow around instead of through the animal*. It is the accumulation of the charges on opposite skin surfaces that further adjusts the electric field in and in the vicinity of the shark by adding ${}^{anl}\mathbf{E}_{anl}$ in the animal and ${}^{str}\mathbf{E}_{anl}$ in its vicinity in the stream. These statements are based on the superposition principle of electric fields.

Thus, in the shark, the ensuing distributed dipole field ${}^{anl}\mathbf{E}_{anl}$ of the + and – charges on opposite skin surfaces adds vectorially to the oppositely directed motional electric field ${}^{anl}\mathbf{E}_{str}$ of the – and + charges at the lateral boundaries of the stream to create a total motional electric field ${}^{anl}\mathbf{E}_{str} + {}^{anl}\mathbf{E}_{anl}$, with ${}^{anl}\mathbf{E}_{str}$ *counteracting*, and ${}^{anl}\mathbf{E}_{anl}$ *aiding* $\mathbf{v} \times \mathbf{B}_v$. Outside the animal, the distributed dipole field ${}^{str}\mathbf{E}_{anl}$ of the charges on opposite skin surfaces adds vectorially to the motional electric field ${}^{str}\mathbf{E}_{str}$ of the charges at the lateral boundaries of the stream to create a total motional electric field ${}^{str}\mathbf{E}_{str} + {}^{str}\mathbf{E}_{anl}$, again with ${}^{str}\mathbf{E}_{str}$ *counteracting*, and ${}^{str}\mathbf{E}_{anl}$ on the average *aiding* the $\mathbf{v} \times \mathbf{B}_v$. Note that ${}^{anl}\mathbf{E}_{anl}$ and ${}^{str}\mathbf{E}_{anl}$ apply to two different regions of one and the same electric field, to wit, the field of the charge accumulations on the animal's opposite skin surfaces. The two regions, moving at the same velocity $\mathbf{v}$, differ in that the animal features a relatively high resistivity and the stream a relatively low resistivity.

Apology! The author has for the sake of simplicity taken the liberty to diagram (Fig. 2.3) the transverse ampullary canals leading to the left and right skin pores as if they emerge from a single median capsule, as do the caudal-rostral and ventro-dorsal ampullary canals. Yet, this is for the transverse ampullae not the real situation, as in the animal the ampullary canals do not cross the sagittal plane. Formally, the configuration should have been as given in the diagram to the right. Nevertheless, since the two medium pores essentially sense the same potential, the two transverse ampullary canals and their capsules in series together sense the left-right pore-to-pore potential differentially at any rate. Actually, it offers the possibility to sense the difference in the electric field at the left and right side of the body.

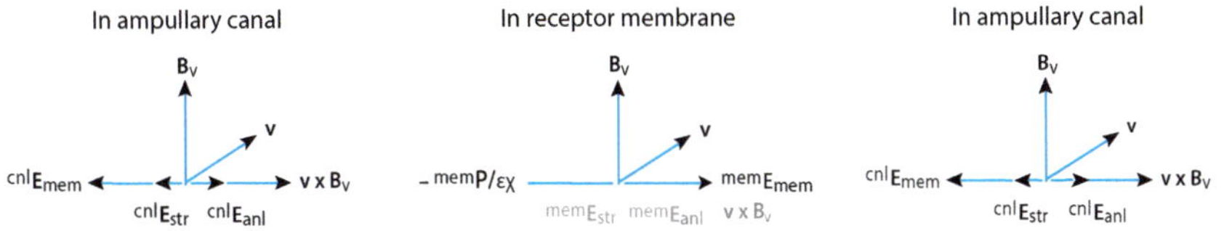

Fig. 2.3 Transverse ampullary canals leading to the left and right skin pores as if they emerge from a single median capsule

Paradoxically, even though the + and – charge accumulations on opposite skin surfaces tend not only to increase the current flowing along the outside past the animal but also the current through the animal, nevertheless the current through the animal is reduced! This is indeed true, since the charge accumulations on the skin are the consequence of the animal being of higher resistivity, on the average ≈ 2 kohm cm, than the seawater, ≈ 20 ohm cm. It is this high resistivity that causes the electric current in the skin and body tissues to be *reduced* by the animal *adding* an *oppositely* directed current of density ${}^{anl}\mathbf{J}_{anl}$ to the current of density ${}^{anl}\mathbf{J}_{str}$ the animal would receive from the stream when the resistivity of the animal was the same as that of the seawater in the stream. The counter current holds despite the *larger* motional electric force resulting from the charge accumulations on opposite skin surfaces. Note that the internal counter current is continuous with the additional current flowing along the outside past the animal, as required by the principle of continuity.

What does this mean for the resistive forces $-\rho^{anl}\mathbf{J}_{str}$ and $-\rho^{anl}\mathbf{J}_{anl}$ inside the animal? Since fundamentally, $\mathbf{v} \times \mathbf{B} - \rho\mathbf{J} + \mathbf{E} = 0$, at constant $\mathbf{v} \times \mathbf{B}$ the resistive force $-\rho\mathbf{J}$ increases when the motional electric field $\mathbf{E}$ decreases due to the + and – charge accumulations on opposite skin surfaces of the animal. Hence, in the animal, the vector sum of $-\rho^{anl}\mathbf{J}_{str}$ and $-\rho^{anl}\mathbf{J}_{anl}$ is *increased* due to the animal's decreased ${}^{anl}\mathbf{E}_{anl}$ and increased resistivity ρ. Conversely, in the stream, the return current of density ${}^{str}\mathbf{J}_{anl}$ and the ${}^{str}\mathbf{J}_{str}$ add in the same direction, allowing the total electric current to flow largely around and only partly through the animal (Fig. 2.2, middle). See also the vector diagrams of Fig. 2.2 and, for further physical clarification, the insight-providing axiom presented in Appendix 6.

Inside the shark, but outside the ampullary sense organs, the motionally induced electric force $\mathbf{v} \times \mathbf{B}_v$, the resistive force $-\rho^{anl}\mathbf{J}_{str}$ due to ${}^{anl}\mathbf{J}_{str}$, the resistive force $-\rho^{anl}\mathbf{J}_{anl}$ due to ${}^{anl}\mathbf{J}_{anl}$, the motional electric force ${}^{anl}\mathbf{E}_{str}$ of the charges at the boundaries of the stream, and the motional electric force ${}^{anl}\mathbf{E}_{anl}$ of the charges on opposite skin surfaces of the animal are in dynamic equilibrium, $\mathbf{v} \times \mathbf{B}_v - \rho^{anl}\mathbf{J}_{str} - \rho^{anl}\mathbf{J}_{anl} + {}^{anl}\mathbf{E}_{str} + {}^{anl}\mathbf{E}_{anl} = 0$, where ρ is the relatively high average resistivity along inside paths through the skin and body tissues (Fig. 2.2, right). Outside the animal, the motionally induced electric force $\mathbf{v} \times \mathbf{B}_v$, the resistive force $-\rho^{str}\mathbf{J}_{str}$, due to ${}^{str}\mathbf{J}_{str}$, the resistive force $-\rho^{str}\mathbf{J}_{anl}$ due to ${}^{str}\mathbf{J}_{anl}$, the ${}^{str}\mathbf{E}_{str}$ of the stream charges at the boundaries of the stream, the ${}^{str}\mathbf{E}_{anl}$ the motional electric force due to the charges on the skin of the animal are in dynamic equilibrium $\mathbf{v} \times \mathbf{B}_v - \rho^{str}\mathbf{J}_{str} - \rho^{str}\mathbf{J}_{anl} + {}^{str}\mathbf{E}_{str} + {}^{str}\mathbf{E}_{anl} = 0$, where ρ is the relatively low resistivity along outside paths (Fig. 2.2, left). The $-\rho^{str}\mathbf{J}_{str}$ is the part of the $-\rho\mathbf{J}$ field resulting from the $\mathbf{v} \times \mathbf{B}_v$ and the ${}^{str}\mathbf{E}_{str}$ due to the charge accumulations at the boundaries of the stream, whereas the $-\rho^{anl}\mathbf{J}_{anl}$ is the part of the $-\rho\mathbf{J}$ field resulting from the $\mathbf{v} \times \mathbf{B}_v$ and the ${}^{anl}\mathbf{E}_{anl}$ due to the charges on the skin. Inside our ellipsoidal, homogenous resistivity model of the shark, not taking the high resistance of the blind ampullary organs into account, $\mathbf{E}_{anl}$ and $-\rho\mathbf{J}_{anl}$ are essentially uniform like the $\mathbf{v} \times \mathbf{B}_v$ and $\mathbf{E}_{str}$ fields. This uniform-field model will be relaxed below, as the ellipsoidal shape and the homogeneous resistivity are mere idealizations of the real animal.

2.1.1.4 In the Ampullary Sense Organs (Fig. 2.3)

In the ampullary sense organs and the capsules in which the ampullae proper are embedded, the motional electric field ${}^{cnl}\mathbf{E}_{str}$ in the ampullary canal, ${}^{mem}\mathbf{E}_{str}$ in the receptor-cell membranes, and ${}^{cap}\mathbf{E}_{str}$ in the capsule, due to the charge accumulations at the boundaries of the stream, opposing the $\mathbf{v} \times \mathbf{B}_v$, and the motional electric field ${}^{cnl}\mathbf{E}_{anl}$ in the ampullary canal, ${}^{mem}\mathbf{E}_{anl}$ in the receptor-cell membrane, ${}^{cap}\mathbf{E}_{anl}$ in the capsule due to the charge accumulations on the opposite skin surfaces of the animal, supporting the $\mathbf{v} \times \mathbf{B}_v$, are just regions of the same fields as those in the shark's body tissues and the surrounding seawater. However, since the thin epithelial membranes lining the gel-filled ampullary canals and blind ampullae proper are considered purely dielectric, the current density ${}^{cnl}\mathbf{J}_{str}+{}^{cnl}\mathbf{J}_{anl}$ in the ampullary canals, driven by the $\mathbf{v} \times \mathbf{B}$, the ${}^{cnl}\mathbf{E}_{str}$, and the ${}^{cnl}\mathbf{E}_{anl}$, ceases to exist after a very short transient current accumulating + and – charges on the thin epithelial walls of the blind ampullary organs, in particular on the receptor-cell membranes of the sensory epithelia lining the ampullae proper. As a result, the electric field ${}^{cnl}\mathbf{E}_{mem}$ of the accumulated receptor-cell membrane charges nulls the combined action of $\mathbf{v} \times \mathbf{B}_v$, ${}^{cnl}\mathbf{E}_{str}$, and ${}^{cnl}\mathbf{E}_{anl}$ in the canal, causing the transient ${}^{cnl}\mathbf{J}_{str}+{}^{cnl}\mathbf{J}_{anl}$ to vanish after a very short time. Hence, in the stationary state, $\mathbf{v} \times \mathbf{B}_v + {}^{cnl}\mathbf{E}_{str} + {}^{cnl}\mathbf{E}_{anl} + {}^{cnl}\mathbf{E}_{mem} = 0$, for now not taking the excitation of the sensory epithelia in the wall of the blind ampullae proper into account. Since the $\mathbf{v} \times \mathbf{B}_v$ and ${}^{cnl}\mathbf{E}_{str}$ in our model are virtually uniform, the ${}^{cnl}\mathbf{E}_{anl}$ and ${}^{cnl}\mathbf{E}_{mem}$ fields, when summed, are in the ampullary canals uniform as well, as a consequence of the self-adjusting charge distribution on the wall of the canals and ampullae proper (Fig. 2.4, left and right).

So far, the epithelial membranes lining the ampullary canals and the blind ampullae proper were assumed to be purely dielectric. In the live animal, though, the lobed ampullae proper contain the sensory maculae, which, being electrically excitable, are not purely dielectric, but feature a *negative resistance*. However,

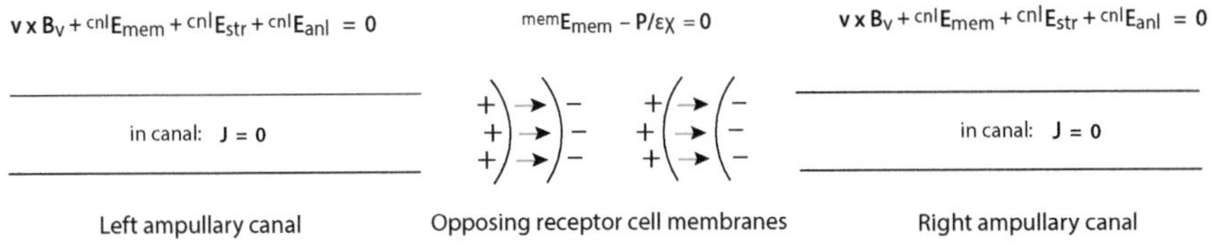

Fig. 2.4 Forces on positive ions in ampullary canals and receptor membranes. Left and right: given that in ampullary canals total current density $\mathbf{J}$ and, hence, total ohmic field $-\rho\mathbf{J}$ are zero, motionally induced electric field $\mathbf{v} \times \mathbf{B}_v$, motional electric field ${}^{cnl}\mathbf{E}_{str}$ of space charges at boundaries of stream, electric field ${}^{cnl}\mathbf{E}_{anl}$ of surface charges on opposite sides of animal, and electric field ${}^{cnl}\mathbf{E}_{mem}$ of surface charges on thin epithelial lining of ampullary canal and ampulla proper are in dynamic equilibrium, $\mathbf{v} \times \mathbf{B}_v + {}^{cnl}\mathbf{E}_{str} + {}^{cnl}\mathbf{E}_{anl} + {}^{cnl}\mathbf{E}_{mem} = 0$. Middle: in sensory membranes of ampullae proper, ${}^{mem}\mathbf{E}_{mem}$, field inside the membrane of charges on opposite sides of receptor membrane, is orders of magnitude stronger than $\mathbf{v} \times \mathbf{B}_v+{}^{cnl}\mathbf{E}_{str}+{}^{cnl}\mathbf{E}_{anl}$, due to thinness of membrane as compared to great length of ampullary canal. Grayed-out quantities negligibly small. Hence, ${}^{mem}\mathbf{E}_{mem}$ force on positive ions, including gating charges of voltage-sensitive ion channels, in receptor membrane for all practical purposes is balanced by $-{}^{mem}\mathbf{P}/\varepsilon\chi_{mem}$. Hence, ${}^{mem}\mathbf{E}_{mem}-{}^{mem}\mathbf{P}/\varepsilon\chi_{mem} = 0$

here we will first consider the membranes of the maculae to be perfect dielectric, though they actively produce negative input currents exiting to the outside through the ampullary canals, in response to pore-positive stimuli, the consequences of which will be detailed in Sect. 2.3.2. Next considering the sensory membranes excitable will change the situation, but not invalidate the physical reasoning. This two-step approach of first considering the purely physical currents and then adding the physiological currents ties together the physical approach with the more recent findings concerning the amplifying physiological operation of the receptor cells.

2.1.1.5 In the Receptor-Cell Membranes (Fig. 2.3)

As the surface charges nulling the ampullary canal currents accumulate on the side of the receptor-cell membranes facing the ampullary gel, necessarily equal amounts of surface charges of opposite polarity accumulate on the other, the capsular-stroma side of the receptor-cell membranes. The membranes thus form ultra-thin, electrolytic capacitors, in which, by geometric design, the electric ${}^{mem}\mathbf{E}_{mem}$ field of the ampullary charges is orders of magnitude stronger than the electric ${}^{cnl}\mathbf{E}_{cnl}$ fields in the conductive, gelatinous contents of the ampullary canals and ampullae proper. Note that the $\mathbf{E}_{mem}$ field and $\mathbf{E}_{cnl}$ field, respectively inside and outside the receptor membrane, refer to two different regions of the same electric field produced by the same charges. The resulting dense concentrations of surface charges subject the ultra-thin receptor membranes to such strong electric fields that the $\mathbf{v} \times \mathbf{B}_v$, ${}^{mem}\mathbf{E}_{str}$, and ${}^{mem}\mathbf{E}_{anl}$ in the membranes are negligibly small compared to the ${}^{mem}\mathbf{E}_{mem}$. Hence, even though the charges on the receptor membranes result from the body-wide $\mathbf{v} \times \mathbf{B}_v$, ${}^{anl}\mathbf{E}_{str}$, and ${}^{anl}\mathbf{E}_{anl}$ fields, within the ultra-thin membranes, it is almost exclusively the orders of magnitude stronger ${}^{mem}\mathbf{E}_{mem}$ field that acts on the bound molecular charges and the gating charges of the voltage-sensitive ion channels.

In the receptor membranes, the bound molecular charges that the geometrically greatly amplified ${}^{mem}\mathbf{E}_{mem}$ field acts upon are held back by the dielectric polarization reactance force $-{}^{mem}\mathbf{P}/\varepsilon\chi$, with $\mathbf{P}$ the polarization density, ε the electric permittivity, and χ the electric susceptibility. Thus, since in the membranes the $\mathbf{v} \times \mathbf{B}_v$, ${}^{mem}\mathbf{E}_{str}$, and ${}^{mem}\mathbf{E}_{anl}$ are vanishingly small compared to the ${}^{mem}\mathbf{E}_{mem}$, the space-averaged ${}^{mem}\mathbf{E}_{mem}$ is for all intents and purposes in dynamic equilibrium with the membrane's molecular reactance force $-{}^{mem}\mathbf{P}/\varepsilon\chi$. Hence, practically, we may reckon ${}^{mem}\mathbf{E}_{mem} - {}^{mem}\mathbf{P}/\varepsilon\chi = \mathbf{0}$. By acting on the gating chargers of the ion channels, these ultimately are the forces responsible for the detection of the electrical signals (Figs. 2.2, right and 2.3, middle). The $-{}^{mem}\mathbf{P}/\varepsilon\chi$ restraining the bound ionic charges in the receptor-cell membranes represents the functional counterpart of the $-\rho^{anl}\mathbf{J}_{anl}$ restraining the ionic charges moving through the animal's body tissues outside the ampullary sense organs. How we arrive at this important conclusion will be explained in the next section.

2.1.2 *Physical Principles of Passive Electro-orientation: In Terms of Voltages*[2]

How strong is the excitation field $\mathbf{E}_{mem}$ in the electroreceptor cell membranes of a shark drifting in an open-ocean stream? To gauge $\mathbf{E}_{mem}$, we will express the electrical forces in terms of voltages and potential differences, that is, *as their line integrals along paths from one point to another*, rather than in terms of dynamic equilibria at single points.

2.1.2.1 Governing Equations

In mathematical physics, electric and magnetic field equations can be expressed in differential form in terms of *forces* or in integral form in terms of *voltages*. The two modes of expression give both penetrating yet different insights, hence they are complementary. It must be well understood that $\mathbf{v} \times \mathbf{B}$ and $-\rho\mathbf{J}$ are gradients of voltages, as they tend to increase the voltage, whereas $\mathbf{E}$ is the *negative* of the gradient of the voltage, as it tends to decrease the voltage. The perilous habit of not consistently adhering to either the force approach or the voltage approach may lead to great confusion. I have struggled enough with the problem to feel urged to present the two approaches side by side, yet each on its own to avoid misapprehensions.

In terms of the *forces* governing the motional electric field a shark drifting in an open-ocean stream receives, the stationary situation is ruled by the equation $\mathbf{v} \times \mathbf{B} - \rho\mathbf{J} + \mathbf{E} = 0$. That is, the driving force $\mathbf{v} \times \mathbf{B}$, the resistive force $-\rho\mathbf{J}$, and the electrostatic force $\mathbf{E}$ of the – and + charges accumulated by the electric current of density $\mathbf{J}$ are in dynamic equilibrium. In the stream, the $\mathbf{v} \times \mathbf{B}$ and $\mathbf{J}$ are in the positive → or ↑ direction in our examples; whereas the $-\rho\mathbf{J}$ and $\mathbf{E}$ are in the negative ← or ↓ direction. When $\mathbf{E}$ approaches nil, $\mathbf{v} \times \mathbf{B}$ and $-\rho\mathbf{J}$ tend to a steady equilibrium, that is, in the limit they are of the same strength but opposite in direction, hence, $\mathbf{v} \times \mathbf{B} - \rho\mathbf{J} = 0$. However, when $\mathbf{E}$ is not nil due to + and – charge accumulations at the lateral boundaries of the stream, $\mathbf{v} \times \mathbf{B} - \rho\mathbf{J} + \mathbf{E} = 0$. Hence, $\mathbf{E}$ causes the strength of $-\rho\mathbf{J}$ to diminish in not perfectly loaded ocean streams. These intricate force-based relationships were the topic of the previous section.

In terms of *voltages*, given that the electric force $\mathbf{E}$ equals the negative of the voltage gradient $\nabla\varphi$, the steady situation is governed by the *integral* equation $\int \mathbf{v} \times \mathbf{B} \cdot \mathbf{ds} + \int -\rho\mathbf{J} \cdot \mathbf{ds} = \int \nabla\varphi \cdot \mathbf{ds}$, that is, the electrical energy imparted $\int \mathbf{v} \times \mathbf{B} \cdot \mathbf{ds}$ plus the negative of the electrical energy dissipated $\int -\rho\mathbf{J} \cdot \mathbf{ds}$ equals the electrical energy $\int \nabla\varphi \cdot \mathbf{ds}$ stored into the electric field of the + and – charges accumulated by the current of density $\mathbf{J}$. When $\int \nabla\varphi \cdot \mathbf{ds}$ approaches nil, the imparted energy tends to equal the energy dissipated, that is, in the limit $\int \mathbf{v} \times \mathbf{B} \cdot \mathbf{ds} + \int -\rho\mathbf{J} \cdot \mathbf{ds} = 0$, rendering the physically adequate stimulus $\int -\rho\mathbf{J} \cdot \mathbf{ds}$ equal to $-\int \mathbf{v} \times \mathbf{B} \cdot \mathbf{ds}$. However,

[2] At this stage, the reader is advised to review the mathematical concepts of *divergence* and *curl*, on which the terms *voltage gradients* and *potential gradients* are founded. See Appendix 3.

when $\int \nabla\varphi \cdot \mathbf{ds}$ is positive, $\int \mathbf{v} \times \mathbf{B} \cdot \mathbf{ds} + \int -\rho \mathbf{J} \cdot \mathbf{ds} = \int \nabla\varphi \cdot \mathbf{ds}$, part of the electrical energy is stored, increasing $\int \nabla\varphi \cdot \mathbf{ds}$ by the separation of – and + charges, thereby reducing the energy dissipated by $\int -\rho\mathbf{J} \cdot \mathbf{ds}$ and rendering the physically adequate stimulus $\int -\rho\mathbf{J} \cdot \mathbf{ds}$ less than $-\int \mathbf{v} \times \mathbf{B} \cdot \mathbf{ds}$, as in not perfectly loaded ocean streams. These energy-based relationships are the topic of the present section.

All very well, but I must reiterate: the limit of zero charge accumulations, hence zero **E**, can never be reached, since the seawater return path does not have zero resistance. Common lack of awareness concerning this fundamental issue has led in the scientific literature to an erroneous, oversimplified version of my rigorous physical analysis, one of the main conundrums to be clarified in Sect. 2.3. As Einstein advocated: "As simple as possible, but not simpler."

Since the **v** and $\mathbf{B}_v$ in the stream, in the drifting shark, and in the still water are—at least for now—assumed to be constant and uniform, hence free of curl, $\mathbf{v} \times \mathbf{B}_v$, $-\rho^{str}\mathbf{J}_{str}$, $-\rho^{str}\mathbf{J}_{anl}$, $-\rho^{anl}\mathbf{J}_{anl}$, $-\rho^{anl}\mathbf{J}_{str}$ and $-\rho^{stw}\mathbf{J}_{str}$, are all *potential gradients*, pointing energetically uphill. However, ${}^{str}\mathbf{E}_{str}$, ${}^{str}\mathbf{E}_{anl}$, ${}^{anl}\mathbf{E}_{anl}$, ${}^{anl}\mathbf{E}_{str}$, and ${}^{stw}\mathbf{E}_{str}$ point energetically downhill, hence inherently are the *negatives of the potential gradients* ${}^{str}\nabla\varphi_{str}$, ${}^{str}\nabla\varphi_{anl}$, ${}^{anl}\nabla\varphi_{anl}$, ${}^{anl}\nabla\varphi_{str}$, and ${}^{stw}\nabla\varphi_{str}$ where ${}^{str}\mathbf{E}_{str} = -{}^{str}\nabla\varphi_{str}$ is the field of the – and + space charges at the lateral boundaries of the stream measured in the stream, ${}^{str}\mathbf{E}_{anl} = -{}^{str}\nabla\varphi_{anl}$ the field of the + and – charges on opposite skin surfaces of the animal measured in the stream, ${}^{stw}\mathbf{E}_{str} = -{}^{stw}\nabla\varphi_{str}$ the field of the space charges at the lateral boundaries of the stream measured along the return path through still deeper water and the ocean bottom, ${}^{anl}\mathbf{E}_{str} = -{}^{anl}\nabla\varphi_{str}$ the field resulting from the space charges at the lateral boundaries of the stream measured in the animal, ${}^{anl}\mathbf{E}_{anl} = -{}^{anl}\nabla\varphi_{anl}$ the field of the charges on opposite skin surfaces of the animal measured in the animal.

Similarly, in ampullary sense organs, the $-\mathbf{P}/\varepsilon\chi$ is a voltage gradient. Furthermore, ${}^{cnl}\mathbf{E}_{str} = -{}^{cnl}\nabla\varphi_{str}$ is the field of the space charges at the lateral boundaries of the stream measured in the ampullary canal, ${}^{cnl}\mathbf{E}_{anl} = -{}^{cnl}\nabla\varphi_{anl}$ the field of the charges on opposite skin surfaces in the ampullary canal measured in the canal, ${}^{cnl}\mathbf{E}_{mem}={}^{cnl}\nabla\varphi_{mem}$ the field of the charges on opposite sides of the receptor-cell membranes measured in the ampullary canal, ${}^{mem}\mathbf{E}_{mem} = -{}^{mem}\nabla\varphi_{mem}$ the field resulting from the charges on opposite sides of the receptor-cell membranes measured inside the receptor-cell membranes. The ρ denotes the resistivity at the point of observation or along the path of integration, either in the ocean or in the animal. The importance of focusing on the fields that are free of curl, hence conservative potential gradients, is the characteristic that their pore-to-pore line integrals are path-independent, keeping mathematics initially uncluttered for the sake of clarity.

At the lateral boundaries of the stream, where $\mathbf{v} \times \mathbf{B}_v$ changes spatially steeply in the *parallel* direction, $\mathbf{v} \times \mathbf{B}_v$ is not free of divergence, that is charges accumulate. At the boundaries of the stream, where $\mathbf{v} \times \mathbf{B}_v$ changes spatially steeply in the *normal* direction, as at the bottom of the stream, $\mathbf{v} \times \mathbf{B}_v$ is not free of curl, hence is a voltage gradient, not a potential gradient. These issues will be discussed later in greater detail, when we will relax the constraints. Until then, we will not cross the boundaries of the stream but stay either entirely within the stream or entirely within the waters outside the stream, where in our model we—for now—assume the divergence and the curl to be zero, and assume the velocity **v** of the water and the animal, the magnetic field **B**, and the resistivity ρ of the water and the animal to be uniform and time-independent.

2.1.2.2 Introduction and Preview of Analysis

To illustrate the governing equations in their simplest form, we will analyze three instances of motional electric measurements with long salt bridges connected by non-polarizable electrodes to an ideal voltmeter, by definition having infinite input resistance, hence not drawing any electric current: (1) the measuring system drifting in an ocean stream, (2) the measurement system moored in an ocean stream, and (3) the measurement system anchored on the bottom below an ocean stream. The salt bridges will be labeled *cnl* as they are the equivalents of the sharks' ampullary canals; the electrodes plus sensing elements will be labeled *mem* as they are the equivalents of the sharks' electroreceptor membranes. The salt bridges and electrodes are assumed to have such slender shape and small volume that they neither significantly alter the motion of the water nor the imposed electric current fields due to electrical charges accumulating on opposite surfaces. The differential pairs of salt-bridge pores will be oriented horizontally for interaction with the vertical component of the earth's magnetic field and vertically for interaction with the horizontal component of the earth's magnetic field.

We will analyze the same three instances for a shark with a pair of voltage-sensitive ampullae of Lorenzini, this time also taking the animal's body in which the ampullae are embedded, into account. The shark's body is large and of quite different resistivity than the surrounding seawater, so that the imposed electric current fields cause significant charge accumulations on opposite sides of the animal, adding a distributed dipole current field. The amount of charge accumulations and their effect on the current field depends on the strength of the electric current in the stream, which in turn depends on the loading of the stream as well as on the average resistivity of the shark relative to the resistivity of the surrounding seawater. As for the effect of the shark's body on the motion of the water, we will estimate the necessary corrections below. Warning: the case of the animal may be difficult to grasp. Rest assured though, the oceanographic case with towed and moored electrodes correctly exhibits the bare physical fundamentals. It is the presence of the shark's resistive body, about a hundred times more resistive than the seawater it displaces, that clutters the analysis with intricate detail.

First, we will preview the general rules governing (1) the electric fields physical oceanographers measure at sea with drifting and moored salt-bridge electrodes connected to a voltmeter, and (2) drifting and stationary sharks measure with their electroreceptors, the ampullae of Lorenzini. We will list the idealized situations considered thus far.

If the water is *still* and devoid of electric fields, *moored* salt-bridge electrodes connected to a voltmeter shall measure nil. Hence, this is a good situation to null the measurement system. Sharks may similarly reset their electroreceptors by not accelerating or decelerating for a time long compared to their time constant of adaptation.

~ ***Electrodes Drifting in Stream*** The stream is *flowing* and by interaction with the earth's magnetic field induces an electric current in the stream, accumulates – and + charges at the lateral boundaries of the stream, and imposes an ohmic resistive force on the ionic charge carriers in the stream. *Drifting* electrodes & voltmeter measure the line integral of the *resistive* $-\rho\mathbf{J}$ between the salt-bridge pores along a path through the seawater, which equals the line integral of $-\mathbf{v} \times \mathbf{B}$ plus the line integral of $\nabla\varphi$:

$$\text{!!! Ergo horizontally}: \int^{\text{mem}} \nabla\varphi_{\text{mem}} \cdot \mathbf{ds} = \int -\rho^{\text{str}} \mathbf{J}_{\text{str}} \cdot \mathbf{ds}$$

~ ***Shark Drifting in Stream*** The stream is *flowing* and by interaction with the earth's magnetic field induces an electric current in the stream, accumulates – and + charges at the lateral boundaries of the stream, and imposes an ohmic resistive force on the ionic charge carriers in the stream. The ampullae of Lorenzini of the *drifting* shark measure the line integral of the *resistive* $-\rho\mathbf{J}$ between the skin pores on opposite sides of the animal along a path through the seawater, which equals the line integral of the $-\rho\mathbf{J}$ drifting electrodes detect *plus* the line integral of an additional $-\rho\mathbf{J}$ resulting from part of the electric current in the stream to flow around the animal instead of through the animal, as the animal is more resistive than the seawater.

$$\text{!!! Ergo}: \int^{\text{mem}} \nabla\varphi_{\text{mem}} \cdot \mathbf{ds} = \int \left(-^{\text{str}} \rho\mathbf{J}_{\text{str}} -^{\text{str}} \rho\mathbf{J}_{\text{anl}} \right) \cdot \mathbf{ds}$$

~ ***Electrodes Moored in Stream*** The stream is *flowing* and by interaction with the earth's magnetic field induces an electric current in the stream, accumulates – and + charges at the lateral boundaries of the stream, and imposes an ohmic resistive force on the ionic charge carriers in the stream. *Moored* electrodes & voltmeter measure the line integral of the *voltage gradient* $\nabla\varphi$ between the salt-bridge pores along a path through the seawater, which equals the line integral of $\mathbf{v} \times \mathbf{B} - \rho\mathbf{J}$:

$$\text{!!! Ergo}: \int^{\text{mem}} \nabla\varphi_{\text{mem}} \cdot \mathbf{ds} = \int^{\text{str}} \nabla\varphi_{\text{str}} \cdot \mathbf{ds}$$

~ ***Shark Stationary in Stream*** The stream is *flowing* and by interaction with the earth's magnetic field induces an electric current in the stream, accumulates – and + charges at the lateral boundaries of the stream, and imposes an ohmic resistive force on the ionic charge carriers in the stream. The ampullae of Lorenzini of the *stationary* shark measures the line integral of the *voltage gradient* $\nabla\varphi$ between the skin pores on opposite sides of the animal along a path through the seawater, which equals the line integral of $\nabla\varphi$ stationary electrodes detect *plus* the line integral of an additional $\nabla\varphi$ due to the charge accumulations on opposite sides of the body, as the shark is more resistive than the seawater.

$$!!!\ \text{Ergo}: \int^{\text{mem}} \nabla\varphi_{\text{mem}} \cdot \mathbf{ds} = \int \left({}^{\text{str}}\nabla\varphi_{\text{str}} + {}^{\text{str}}\nabla\varphi_{\text{anl}} \right) \cdot \mathbf{ds}$$

~ *Electrodes Moored on Bottom Under Stream* The return volume of water is *still* and features an electric field of voltage gradient $\nabla\varphi$, produced by the – and + charge accumulations at the lateral boundaries of the stream. *Moored* electrodes & voltmeter measure the line integral of the *voltage gradient* $\nabla\varphi$ between the salt-bridge pores along a path through the seawater, which equals the line integral of $-\rho\mathbf{J}$:

$$!!!\ \text{Ergo}: \int^{\text{mem}} \nabla\varphi_{\text{mem}} \cdot \mathbf{ds} = \int^{\text{stw}} \nabla\varphi_{\text{stw}} \cdot \mathbf{ds}$$

~ *Shark Resting on Bottom, Under Stream* The return volume of water is *still* and features an electric field of voltage gradient $\nabla\varphi$, produced by the charge accumulations at the lateral boundaries of the stream. The ampullae of Lorenzini of the resting shark measures the line integral of the *voltage gradient* $\nabla\varphi$, which equals the line integral of $\nabla\varphi$ moored electrodes detect *plus* the line integral of an additional $\nabla\varphi$ due to the charge accumulations on opposite sides of the body, as the shark is more resistive than the seawater.

$$!!!\ \text{Ergo}: \int^{\text{mem}} \nabla\varphi_{\text{mem}} \cdot \mathbf{ds} = \int \left({}^{\text{stw}}\nabla\varphi_{\text{str}} + {}^{\text{stw}}\nabla\varphi_{\text{anl}} \right) \cdot \mathbf{ds}$$

In the following mathematical derivations, it is repeatedly stated that:

"Practically $-{}^{\text{mem}}\mathbf{P}_{\text{mem}}/\varepsilon\chi \cdot \mathbf{ds} = \int^{\text{mem}}\nabla\varphi_{\text{mem}} \cdot \mathbf{ds}$ and $\int({}^{\text{cnl}}\nabla\varphi_{\text{mem}} + {}^{\text{mem}}\nabla\varphi_{\text{mem}}) \cdot \mathbf{ds} = 0$."

$\int^{\text{mem}}\nabla\varphi_{\text{mem}} \cdot \mathbf{ds} = -{}^{\text{mem}}\mathbf{P}_{\text{mem}}/\varepsilon\chi \cdot \mathbf{ds}$ is virtually true since the $\int^{\text{mem}}\nabla\varphi_{\text{mem}}$ inside the sensory membrane is many orders of magnitude stronger than any other voltage gradients acting from the outside on the sensory membrane. $\int({}^{\text{cnl}}\nabla\varphi_{\text{mem}} + {}^{\text{mem}}\nabla\varphi_{\text{mem}}) \cdot \mathbf{ds} = 0$ is the very principle of voltmeters with their test leads, including the ampullae of Lorenzini with their gel-filled canals. Let us review why:

It must be well understood that the electric charges nulling the electric current in the salt bridges and in the ampullary canals are the same charges as those producing the electric fields in the voltage-sensitive receptor-cell membranes. The charges on the wall of the ampullary canal shaping the electric field in the canal do not enter the pore-to-pore line integral and need not be discussed any further. Hence, the salt-bridge and ampullary canal pore-to-pore integral of ${}^{\text{cnl}}\nabla\varphi_{\text{mem}} + {}^{\text{mem}}\nabla\varphi_{\text{mem}}$ is nil. That is, the electrical charges on either side of the receptor membranes build up until the electric fields in the salt bridges or ampullary canals are nulled, and the voltages between the pores appear across the receptor membranes. That is the virtue for a true voltmeter, not drawing any current nor adding any voltage to the circuit, *unless it is moving in a magnetic field*, which adds a voltage $\int\mathbf{v} \times \mathbf{B}_{\text{v}} \cdot \mathbf{ds}$. Actually, the ampullae of Lorenzini in response actively produce a negative electric input current, to be discussed below but here not yet taken into account.

The stream flows positive from west to east →, the magnetic field points positive up ↑ assuming we are in the southern hemisphere. The equations in the following mathematical discourse describe the various situations as they are observed in the earth frame of reference. How the equations transform to the shark frame of reference will be explained below.

The mathematical details of the paragraphs marked with an "approximate" sign (~) may be glanced over, only to read the concluding lines marked with three exclamation marks (!!!) and with a pound sign (#). Warning, to envision the various situations is challenging, especially when we introduce the shark into the stream. So that each situation can stand on its own, the in-between steps are spelled out every time again. The readers who will understand the following situations will find out that it all is rather simple in hindsight. Sharks leisurely practice the intricate physics.

Superscripts are omitted where it is obvious from the context where the pertinent physical quantities are measured.

The situations concerning electric fields measured technically with electrodes and by sharks with the ampullary system are self-contained and can be studied in isolation or in arbitrary order.

2.1.2.3 Detailed Mathematical Analysis

~ Electrodes Drifting in Stream, Interaction with $\mathbf{B}_v$, (a) Electric Field Horizontal (Fig. 2.1)

In stream:

$$\mathbf{v} \times \mathbf{B}_v - \rho^{str}\mathbf{J}_{str} + {}^{str}E_{str} = 0 \qquad \text{forces on unit}^{+}\text{ions}$$

$$\mathbf{v} \times \mathbf{B}_v - \rho^{str}\mathbf{J}_{str} = {}^{str}\nabla\varphi_{str} \qquad \text{potential gradients}$$

Salt bridge pore-to-pore line integral along paths through flowing seawater:

$$\text{Path 1:}\quad V_{p2} - V_{p1} = \int\left(\mathbf{v} \times \mathbf{B}_v - \rho^{str}\mathbf{J}_{str}\right)\cdot d\mathbf{s} = \int^{str}\nabla\varphi_{str}\cdot d\mathbf{s} \qquad \textit{electric current}$$

In drifting salt bridges and sensing elements, respectively: *no electric current*!

$$\mathbf{v} \times \mathbf{B}_v + {}^{cnl}E_{str} + {}^{cnl}E_{mem} = 0 \text{ and } -{}^{mem}\mathbf{P}_{mem}/\varepsilon\chi + {}^{mem}E_{mem} = 0 \quad \text{forces on unit}^{+}\text{ions}$$

$$\mathbf{v} \times \mathbf{B}_v = {}^{cnl}\nabla\varphi_{str} + {}^{cnl}\nabla\varphi_{mem} \text{ and } -{}^{mem}\mathbf{P}_{mem}/\varepsilon\chi = {}^{mem}\nabla\varphi_{mem} \qquad \text{potential gradients}$$

Salt bridge pore-to-pore line integral along paths through salt bridges and sensing elements:

$$\text{Path 2: } V_{p2} - V_{p1} = \int \left(\mathbf{v} \times \mathbf{B}_{v} - {}^{mem}\mathbf{P}_{mem} / \varepsilon\chi \right) \cdot \mathbf{ds}$$
$$= \int \left({}^{cnl}\nabla \varphi_{str} + {}^{cnl}\nabla \varphi_{mem} + {}^{mem}\nabla \varphi_{mem} \right) \cdot \mathbf{ds}$$

Or, since practically $\int -{}^{mem}\mathbf{P}_{mem}/\varepsilon\chi \cdot \mathbf{ds} = \int {}^{mem}\nabla\varphi_{mem} \cdot \mathbf{ds}$ and $\int({}^{cnl}\nabla\varphi_{mem} + {}^{mem}\nabla\varphi_{mem}) \cdot \mathbf{ds} = 0$

$$\text{Path 2: } V_{p2} - V_{p1} = \int \left(\mathbf{v} \times \mathbf{B}_{v} + {}^{mem}\nabla \varphi_{mem} \right) \cdot \mathbf{ds} = \int^{cnl} \nabla \varphi_{str} \cdot \mathbf{ds}$$

From path 2 and path 1 equations:

$$\int \left(\mathbf{v} \times \mathbf{B}_{v} + {}^{mem}\nabla \varphi_{mem} \right) \cdot \mathbf{ds} = \int \left(\mathbf{v} \times \mathbf{B}_{v} - \rho^{str}\mathbf{J}_{str} \right) \cdot \mathbf{ds} = \int^{str} \nabla \varphi_{str} \cdot \mathbf{ds}$$

!!! Ergo horizontally: $\int^{mem} \nabla \varphi_{mem} \cdot \mathbf{ds} = \int -\rho^{str}\mathbf{J}_{str} \cdot \mathbf{ds}$ $\quad\leftarrow$ in negative direction

In brief: the pore-to-pore paths of integration through the salt bridges & sensing elements and the pore-to-pore paths through the seawater have the line integrals of $\mathbf{v} \times \mathbf{B}_v$ and $\nabla\varphi_{str}$ in common, leaving the line integrals of ${}^{mem}\nabla\varphi_{mem} = -\rho^{str}\mathbf{J}_{str} \leftarrow$ for the voltmeter to detect.

Horizontally oriented drifting electrodes normal to the stream thus measure—by interaction with the vertical component of the earth's magnetic field—a horizontally directed ohmic field $-\rho^{str}\mathbf{J}_{str} \leftarrow$ resulting from the motionally induced electric current of density ${}^{str}\mathbf{J}_{str} \rightarrow$ in the stream. The $-\rho^{str}\mathbf{J}_{str}$ is directed normal to the stream and independent of the magnetic heading of the stream.

The $-\rho^{str}\mathbf{J}_{str}$ field becomes stronger, the more heavily the stream is electrically loaded, hence, the smaller the charge accumulations at the boundaries of the stream are, that is, the smaller ${}^{str}\nabla\varphi_{str}$ is and the more is left for the $-\rho^{str}\mathbf{J}_{str}$ to counteract $\mathbf{v} \times \mathbf{B}_v$. In the virtual limit of an electrically shorted stream, $-\rho^{str}\mathbf{J}_{str} = -\mathbf{v} \times \mathbf{B}_v$. Conversely, if the stream flows in an insulating channel, not at all loading the stream, ${}^{str}\nabla\varphi_{str} = \mathbf{v} \times \mathbf{B}_v$, the $-\rho^{str}\mathbf{J}_{str}$ is nil, and the electrodes measure naught. Moreover, the horizontal $-\rho^{str}\mathbf{J}_{str}$ is the *weaker* the closer the stream flows near the magnetic equator.

~ Electrodes Drifting in Stream, Interaction with $\mathbf{B}_h$, (b) Electric Field Vertical

Interaction of the stream and the electrodes with $\mathbf{B}_h$ leads to a vertically directed $\mathbf{v} \times \mathbf{B}_h$. The mathematic analysis is analogous to the above under interaction with $\mathbf{B}_v$.

!!! Ergo vertically: $\int^{mem} \nabla \varphi_{mem} \cdot \mathbf{ds} = \int -\rho^{str}\mathbf{J}_{str} \cdot \mathbf{ds}$ $\quad\uparrow$ or $\downarrow$ direction

Vertically oriented drifting electrodes thus measure—by interaction with the horizontal component of the earth's magnetic field—in the stream a vertically directed ohmic field $-\rho^{str}\mathbf{J}_{str}$ resulting from the motionally induced electric current of density ${}^{str}\mathbf{J}_{str}$. The $-\rho^{str}\mathbf{J}_{str}$ depends on the magnetic heading of the stream: maximally negative ↓ when the stream flows due east, maximally positive ↑ when the streams flows due west, and zero when the stream flows due north or south.

The great difference between the interaction with $\mathbf{B}_h$ and the interaction with $\mathbf{B}_v$ is that, due to the seawater-air interface, wind-driven surface streams are hardly loaded over most of their widths, hence the vertical $-\rho^{str}\mathbf{J}_{str}$ tends to be weak, except near the lateral boundaries of the stream. The vertical $-\rho^{str}\mathbf{J}_{str}$ is the *stronger* the closer the stream flows near or at the magnetic equator.

~ Shark Drifting in Stream, Interaction with $\mathbf{B}_v$, (a) Electric Field Horizontal (Fig. 2.1)

The sensory system enabling *drifting* sharks to detect the motional electric fields of ocean streams—consisting of sensory epithelia, conductive gel-filled ampullary canals, and skin pores—and the oceanographic system to measure the electric fields of ocean streams—consisting of *drifting* voltmeters, electrodes, and open-ended agar-filled salt bridge tubes—are very similar. Yet, there is an important difference: in marine sharks the sensory system is embedded in a body of higher resistivity than the seawater, whereas the oceanographic measurement system is directly exposed to the seawater. The skin and body tissues of the sharks make the analysis much more complicated.

Understand well and keep in mind, by introducing the more resistive shark into the less resistive seawater of the stream, the motionally induced electric current of density ${}^{str}\mathbf{J}_{str}$ flowing across the stream causes + and – electrical charges to accumulate on opposite sides of the animal. They are the charges that force part of the originally undisturbed electric current in the stream to flow around the more than seawater resistive animal instead of flowing through it.

We shall first determine (1) how the ampullary system detects the field inside the drifting shark and then (2) how the field inside the drifting shark relates to the undisturbed field in the ocean stream, and (3) how the signals the drifting shark receives differ from the signals the drifting electrodes detect. The following analysis is tedious. Yet, the results are surprisingly simple, almost self-evident: *Drifting sharks detect the* $-\rho^{str}\mathbf{J}_{str}$ *created by the induced electric current in the stream,—like drifting electrodes do—but enhanced (!) by the* $-\rho^{str}\mathbf{J}_{anl}$ *due to the change in the field pattern the shark creates in the stream's electric current field by being more resistive than the seawater.* So, it may be easier just to trust the results rather than to follow the detailed analysis. Nevertheless, the author wanted for himself formally to ratify his intuition, instead of dodging the intricate details. Sharks are not aware of the science but just practice it leisurely, rendering the author humiliated and envious.

(i) Inside the drifting shark, but outside ampullary sense organs:

$$\mathbf{v}\times\mathbf{B}_{\mathrm{v}} - \rho^{\mathrm{anl}}\mathbf{J}_{\mathrm{str}} - \rho^{\mathrm{anl}}\mathbf{J}_{\mathrm{anl}} + {}^{\mathrm{anl}}E_{\mathrm{str}} + {}^{\mathrm{anl}}E_{\mathrm{anl}} = 0 \qquad \text{forces on unit}^{+}\text{ions}$$

$$\mathbf{v}\times\mathbf{B}_{\mathrm{v}} - \rho^{\mathrm{anl}}\mathbf{J}_{\mathrm{str}} - \rho^{\mathrm{anl}}\mathbf{J}_{\mathrm{anl}} = {}^{\mathrm{anl}}\nabla\varphi_{\mathrm{str}} + {}^{\mathrm{anl}}\nabla\varphi_{\mathrm{anl}} \qquad \text{potential gradients}$$

The ${}^{\mathrm{anl}}\mathbf{E}_{\mathrm{str}} = -{}^{\mathrm{anl}}\nabla\varphi_{\mathrm{str}} \leftarrow$ pertains to the motional electric field in the animal due to the – and + charges at the lateral boundaries of the stream. The ${}^{\mathrm{anl}}\mathbf{E}_{\mathrm{str}}$ opposes the $\mathbf{v} \times \mathbf{B}_{\mathrm{v}}$. The ${}^{\mathrm{anl}}\mathbf{E}_{\mathrm{anl}} = -{}^{\mathrm{anl}}\nabla\varphi_{\mathrm{anl}} \rightarrow$ pertains to the motional electric field in the animal due to the + and – charges on opposite skin surfaces of the animal. The ${}^{\mathrm{anl}}\mathbf{E}_{\mathrm{anl}}$ aids the $\mathbf{v} \times \mathbf{B}_{\mathrm{v}}$.

The $-\rho^{\mathrm{anl}}\mathbf{J}_{\mathrm{str}} \leftarrow$ is the resistive field in the animal due to the electric current of density ${}^{\mathrm{anl}}\mathbf{J}_{\mathrm{str}} \rightarrow$ resulting from the $\mathbf{v} \times \mathbf{B}_{\mathrm{v}}$ in the stream and the electrical charges at the lateral boundaries of the stream, not taking the charges on opposite sides of the animal into account, with ρ the resistivity of the animal. The $-\rho^{\mathrm{anl}}\mathbf{J}_{\mathrm{anl}} \rightarrow$ is the resistive field in the animal due to the electric current of density ${}^{\mathrm{anl}}\mathbf{J}_{\mathrm{anl}} \leftarrow$ resulting from the $\mathbf{v} \times \mathbf{B}_{\mathrm{v}}$ in the animal and the electrical charges on its opposite sides of the animal, not taking the charges on the lateral boundaries of the stream into account, with ρ the resistivity of the animal. Note that ${}^{\mathrm{anl}}\mathbf{J}_{\mathrm{anl}} \leftarrow$ *subtracts* from the ${}^{\mathrm{anl}}\mathbf{J}_{\mathrm{str}} \rightarrow$, as it represents the part of the current that does not flow through, but around the animal. For the effect of a resistive sphere on a uniform electric current field, refer to the axiom introduced in Appendix 6.

Pore-to-pore line integral along paths through skin & body tissues:

$$\text{Path 1:}\quad V_{\mathrm{p2}} - V_{\mathrm{p1}} = \int \left(\mathbf{v}\times\mathbf{B}_{\mathrm{v}} - \rho^{\mathrm{anl}}\mathbf{J}_{\mathrm{str}} - \rho^{\mathrm{anl}}\mathbf{J}_{\mathrm{anl}}\right)\cdot d\mathbf{s} = \int \left({}^{\mathrm{anl}}\nabla\varphi_{\mathrm{str}} + {}^{\mathrm{anl}}\nabla\varphi_{\mathrm{anl}}\right)\cdot d\mathbf{s}$$

In ampullary canals and in receptor-cell membranes, respectively: *no electric current*!

$$\mathbf{v}\times\mathbf{B}_{\mathrm{v}} + {}^{\mathrm{cnl}}E_{\mathrm{str}} + {}^{\mathrm{cnl}}E_{\mathrm{anl}} + {}^{\mathrm{cnl}}E_{\mathrm{mem}} = 0$$
$$\text{and } -{}^{\mathrm{mem}}\mathbf{P}_{\mathrm{mem}}/\varepsilon\chi + {}^{\mathrm{mem}}E_{\mathrm{mem}} = 0 \qquad \text{forces on unit}^{+}\text{ions}$$

$$\mathbf{v}\times\mathbf{B}_{\mathrm{v}} = {}^{\mathrm{cnl}}\nabla\varphi_{\mathrm{str}} + {}^{\mathrm{cnl}}\nabla\varphi_{\mathrm{anl}} + {}^{\mathrm{cnl}}\nabla\varphi_{\mathrm{mem}}$$
$$\text{and } -{}^{\mathrm{mem}}\mathbf{P}_{\mathrm{mem}}/\varepsilon\chi = {}^{\mathrm{mem}}\nabla\varphi_{\mathrm{mem}} \qquad \text{potential gradients}$$

The ${}^{\mathrm{cnl}}\mathbf{E}_{\mathrm{str}} = -{}^{\mathrm{cnl}}\nabla\varphi_{\mathrm{str}} \leftarrow$ pertains to the motional electric field in the ampullary canal due to the – and + charges at the lateral boundaries of the stream. The ${}^{\mathrm{cnl}}\mathbf{E}_{\mathrm{anl}} = {}^{\mathrm{cnl}}\nabla\varphi_{\mathrm{anl}} \rightarrow$ pertains to the motional electric field in the ampullary canal due to the + and – charges on opposite skin surfaces of the animal. The ${}^{\mathrm{cnl}}\mathbf{E}_{\mathrm{mem}} = -{}^{\mathrm{cnl}}\nabla\varphi_{\mathrm{mem}}$ $\leftarrow$ pertains to the motional electric field in the ampullary canal due to the + and – charges on opposite surfaces of the sensory epithelium. The ${}^{\mathrm{mem}}\mathbf{E}_{\mathrm{mem}} = -{}^{\mathrm{mem}}\nabla\varphi_{\mathrm{mem}} \rightarrow$ pertains to the motional electric field in the sensory epithelium due to the + and – charges on opposite skin surfaces of the sensory epithelium.

Pore-to-pore line integral along paths through sense organs:

$$\text{Path 2: } V_{p2} - V_{p1} = \int \left(\mathbf{v} \times \mathbf{B}_v - {}^{mem}\mathbf{P}_{mem} / \varepsilon\chi \right) \cdot \mathbf{ds}$$
$$= \int \left({}^{cnl}\nabla\varphi_{str} + {}^{cnl}\nabla\varphi_{anl} + {}^{cnl}\nabla\varphi_{mem} + {}^{mem}\nabla\varphi_{mem} \right) \cdot \mathbf{ds}$$

Or, since practically $\int -{}^{mem}\mathbf{P}_{mem}/\varepsilon\chi \cdot \mathbf{ds} = \int \nabla\varphi_{mem} \cdot \mathbf{ds}$ and $\int({}^{cnl}\nabla\varphi_{mem} + {}^{mem}\nabla\varphi_{mem}) \cdot \mathbf{ds} = 0$

$$\text{Path 2: } V_{p2} - V_{p1} = \int \left(\mathbf{v} \times \mathbf{B}_v + {}^{mem}\nabla\varphi_{mem} \right) \cdot \mathbf{ds} = \int \left({}^{cnl}\nabla\varphi_{str} + {}^{cnl}\nabla\varphi_{anl} \right) \cdot \mathbf{ds}$$

Comparing path 2 equation with path 1 equation:

$$\int \left(\mathbf{v} \times \mathbf{B}_v + {}^{mem}\nabla\varphi_{mem} \right) \cdot \mathbf{ds} = \int \left(\mathbf{v} \times \mathbf{B}_v - \rho^{anl}\mathbf{J}_{str} - \rho^{anl}\mathbf{J}_{anl} \right) \cdot \mathbf{ds}$$
$$= \int \left({}^{anl}\nabla\varphi_{str} + {}^{anl}\nabla\varphi_{anl} \right) \cdot \mathbf{ds}$$

$$\text{!!! Ergo horizontally: } \int {}^{mem}\nabla\varphi_{mem} \cdot \mathbf{ds}$$
$$= \int \left(-\rho^{anl}\mathbf{J}_{str} - \rho^{anl}\mathbf{J}_{anl} \right) \cdot \mathbf{ds} \qquad \leftarrow \text{in negative direction}$$

In brief: the pore-to-pore paths of integration through the ampullary canals & receptor cell membranes and the pore-to-pore paths through the skin & body tissues have the line integral of $\mathbf{v} \times \mathbf{B}_v$ and the line integral of ${}^{anl}\nabla\varphi_{str} + {}^{anl}\nabla\varphi_{anl}$ due to the charges at the lateral boundaries of the stream and to the charges accumulating on opposite sides of the shark's body in common, leaving the line integrals of ${}^{mem}\nabla\varphi_{mem} = -\rho^{anl}\mathbf{J}_{str} - \rho^{anl}\mathbf{J}_{anl}$ for the ampullary sense organs of the drifting shark to detect.

In the drifting shark, horizontal ampullary pairs normal to the stream thus measure the ohmic $-\rho^{anl}\mathbf{J}_{str} - \rho^{anl}\mathbf{J}_{anl}$, which is horizontally directed ←, normal to the stream, and independent of the stream's magnetic heading.

Note that the pore-to-pore line integrals of $-\rho^{anl}\mathbf{J}_{str}$ and $-\rho^{anl}\mathbf{J}_{anl}$ are of opposite sign, + and – respectively. The positive $-\rho^{anl}\mathbf{J}_{str}$ tends to override the negative $-\rho^{anl}\mathbf{J}_{anl}$ when the stream is lightly loaded, that is, when the charge accumulations at the boundaries of the stream are large, hence ${}^{str}\mathbf{J}_{str}$ is small. Conversely, $-\rho^{anl}\mathbf{J}_{anl}$ tends to override the $-\rho^{anl}\mathbf{J}_{str}$ when the stream is heavily loaded, that is, the charge accumulations at the boundaries of the stream are small, hence ${}^{str}\mathbf{J}_{str}$ is large. Hence, when the stream is heavily loaded, $-\rho^{anl}\mathbf{J}_{anl}$ may even reverse the total line integral of $-\rho^{anl}\mathbf{J}_{str} - \rho^{anl}\mathbf{J}_{anl}$ the animal receives. The $-\rho^{anl}\mathbf{J}_{anl}$ is the stronger, the more resistive the animal is than the seawater.

(ii) This brings us to the Holy Grail question: how does the $-\rho^{anl}\mathbf{J}_{str} - \rho^{anl}\mathbf{J}_{anl}$ in the drifting shark relate to the $-\rho^{str}\mathbf{J}_{str}$ in the undisturbed stream? Not a trivial question, since we know that in the stream and in the shark the ρ, $\mathbf{J}_{anl}$, and $\mathbf{J}_{str}$ have very different values.

In stream:

$$\mathbf{v} \times \mathbf{B}_v - \rho^{str}\mathbf{J}_{str} - \rho^{str}\mathbf{J}_{anl} + {}^{str}\mathrm{E}_{str} + {}^{str}\mathrm{E}_{anl} = 0 \qquad \text{forces on unit}^{+}\text{ions}$$

$$\mathbf{v}\times\mathbf{B}_\mathrm{v} - \rho^{\mathrm{str}}\mathbf{J}_{\mathrm{str}} - \rho^{\mathrm{str}}\mathbf{J}_{\mathrm{anl}} = {}^{\mathrm{str}}\nabla\varphi_{\mathrm{str}} + {}^{\mathrm{str}}\nabla\varphi_{\mathrm{anl}} \qquad \text{potential gradients}$$

The ${}^{\mathrm{str}}\mathbf{E}_{\mathrm{str}} = -{}^{\mathrm{str}}\nabla\varphi_{\mathrm{str}} \leftarrow$ pertains to the motional electric field in the stream due to the – and + charges at the lateral boundaries of the stream. The ${}^{\mathrm{str}}\mathbf{E}_{\mathrm{str}}$ opposes the $\mathbf{v} \times \mathbf{B}_\mathrm{v}$. The ${}^{\mathrm{str}}\mathbf{E}_{\mathrm{anl}} = -{}^{\mathrm{str}}\nabla\varphi_{\mathrm{anl}} \rightarrow$ pertains to the motional electric field in the stream due to the + and – charges on opposite skin surfaces of the animal. The ${}^{\mathrm{str}}\mathbf{E}_{\mathrm{anl}}$ aids the $\mathbf{v} \times \mathbf{B}_\mathrm{v}$. The $-\rho^{\mathrm{str}}\mathbf{J}_{\mathrm{str}} \leftarrow$ pertains to the $\mathbf{v} \times \mathbf{B}_\mathrm{v}$ and the resistive field in the stream due to the – and + charges at the lateral boundaries of the stream, not taking the charges on opposite sides of the animal into account, with ρ the resistivity of seawater. The $-\rho^{\mathrm{str}}\mathbf{J}_{\mathrm{anl}} \rightarrow$ pertains to the $\mathbf{v} \times \mathbf{B}_\mathrm{v}$ and the resistive field in the stream resulting from the charges on opposite sides of the animal, not taking the charges on the lateral boundaries of the stream into account, with ρ the resistivity of the stream. Note that ${}^{\mathrm{str}}\mathbf{J}_{\mathrm{anl}} \rightarrow$ *adds* to the ${}^{\mathrm{str}}\mathbf{J}_{\mathrm{str}} \rightarrow$, as it represents the part of the current that does not flow through, but around the animal, whereas the ${}^{\mathrm{anl}}\mathbf{J}_{\mathrm{anl}} \leftarrow$ subtracts from the ${}^{\mathrm{anl}}\mathbf{J}_{\mathrm{str}} \rightarrow$.

Pore-to-pore line integral along paths through flowing seawater:

$$\text{Path 3:}\quad V_{\mathrm{p2}} - V_{\mathrm{p1}} = \int\left(\mathbf{v}\times\mathbf{B}_\mathrm{v} - \rho^{\mathrm{str}}\mathbf{J}_{\mathrm{str}} - \rho^{\mathrm{str}}\mathbf{J}_{\mathrm{anl}}\right)\cdot\mathbf{ds} = \int\left({}^{\mathrm{str}}\nabla\varphi_{\mathrm{str}} + {}^{\mathrm{str}}\nabla\varphi_{\mathrm{anl}}\right)\cdot\mathbf{ds}$$

In animal:

$$\mathbf{v}\times\mathbf{B}_\mathrm{v} - \rho^{\mathrm{anl}}\mathbf{J}_{\mathrm{str}} - \rho^{\mathrm{anl}}\mathbf{J}_{\mathrm{anl}} + {}^{\mathrm{anl}}E_{\mathrm{str}} + {}^{\mathrm{anl}}E_{\mathrm{anl}} = 0 \qquad \text{forces on unit}^{+}\text{ions}$$

$$\mathbf{v}\times\mathbf{B}_\mathrm{v} - \rho^{\mathrm{anl}}\mathbf{J}_{\mathrm{str}} - \rho^{\mathrm{anl}}\mathbf{J}_{\mathrm{anl}} = {}^{\mathrm{anl}}\nabla\varphi_{\mathrm{str}} + {}^{\mathrm{anl}}\nabla\varphi_{\mathrm{anl}} \qquad \text{potential gradients}$$

Pore-to-pore line integral along paths through skin & body tissues:

$$\text{Path 1:}\quad V_{\mathrm{p2}} - V_{\mathrm{p1}} = \int\left(\mathbf{v}\times\mathbf{B}_\mathrm{v} - \rho^{\mathrm{anl}}\mathbf{J}_{\mathrm{str}} - \rho^{\mathrm{anl}}\mathbf{J}_{\mathrm{anl}}\right)\cdot\mathbf{ds} = \int\left({}^{\mathrm{anl}}\nabla\varphi_{\mathrm{str}} + {}^{\mathrm{anl}}\nabla\varphi_{\mathrm{anl}}\right)\cdot\mathbf{ds}$$

Comparing path 3 equation with path 1 equation: the pore-to-pore line integral of $-\rho^{\mathrm{anl}}\mathbf{J}_{\mathrm{str}} - \rho^{\mathrm{anl}}\mathbf{J}_{\mathrm{anl}}$ in the animal equals the pore-to-pore line integral of $-\rho^{\mathrm{str}}\mathbf{J}_{\mathrm{str}} - \rho^{\mathrm{str}}\mathbf{J}_{\mathrm{anl}}$ in the stream, even though the ρ, the $\mathbf{J}_{\mathrm{str}}$, and the $\mathbf{J}_{\mathrm{anl}}$ have very different values in the animal and in the stream.

$$\text{!!! Ergo horizontally:} \int {}^{\mathrm{mem}}\nabla\varphi_{\mathrm{mem}}\cdot\mathbf{ds}$$
$$= \int\left(-{}^{\mathrm{str}}\rho\mathbf{J}_{\mathrm{str}} - {}^{\mathrm{str}}\rho\mathbf{J}_{\mathrm{anl}}\right)\cdot\mathbf{ds} \qquad \leftarrow \text{in negative direction}$$

In brief: expressed in terms of internal skin & body tissue fields, the sense organs of the drifting shark detect the line integrals of ${}^{\mathrm{mem}}\nabla\varphi_{\mathrm{mem}} = -\rho^{\mathrm{anl}}\mathbf{J}_{\mathrm{str}} - \rho^{\mathrm{anl}}\mathbf{J}_{\mathrm{anl}}$; expressed in terms of the adjacent seawater field, they detect the line integrals of ${}^{\mathrm{mem}}\nabla\varphi_{\mathrm{mem}} = -\rho^{\mathrm{str}}\mathbf{J}_{\mathrm{str}} - \rho^{\mathrm{str}}\mathbf{J}_{\mathrm{anl}}$.

In the drifting shark, horizontal ampullary pairs normal to the stream thus measure, by interaction with the vertical component of the earth's magnetic field, the ohmic field $-\rho^{\mathrm{str}}\mathbf{J}_{\mathrm{str}} - \rho^{\mathrm{str}}\mathbf{J}_{\mathrm{anl}}$, which is horizontally ← directed, normal to the stream, and independent of the stream's magnetic heading.

(iii) How much the $-\rho^{str}\mathbf{J}_{str} - \rho^{str}\mathbf{J}_{anl}$ is *larger*—in animals more resistive than the surrounding medium—than the $-\rho^{str}\mathbf{J}_{str}$ the drifting electrodes detect depends on the resistivity of the seawater stream, the average resistivity of the animal's skin and body tissues, and the three-dimensional shape of the animal. Even though the resistivities and the current densities can be very different in the animal and the stream, mathematical analysis shows that the $-\rho^{str}\mathbf{J}_{anl}$ adds only a modest fraction to the $-\rho^{str}\mathbf{J}_{str}$, nominally say 0.25 times $-\rho^{str}\mathbf{J}_{str}$ for a shark drifting in an ocean stream. The true fraction shall be determined by experiment along the three mutually orthogonal body axes, as it depends on the shape and resistance of the shark under investigation, to be explained below.

Thus, concerning the physical feature of the stream the animal detects:

$$\text{!!! Ergo horizontally}: \int^{mem} \nabla\varphi_{mem} \cdot \mathbf{ds}$$
$$= \approx 1.25 \times \int -\rho^{str}\mathbf{J}_{str} \cdot \mathbf{ds} \qquad \leftarrow \text{in negative direction}$$

! ! ! Physically Adequate Stimulus

We thus arrive at the conclusion that, for a shark drifting in an open ocean stream, the horizontal $\int^{mem}\nabla\varphi_{mem} \cdot \mathbf{ds}$ equals $\int -\rho^{str}\mathbf{J}_{str} \cdot \mathbf{ds} + \int -\rho^{str}\mathbf{J}_{anl} \cdot \mathbf{ds}$. Hence, *the pore-to-pore line integral* $\int -\rho^{str}\mathbf{J}_{str} \cdot \mathbf{ds}$ *of the resistive potential gradient due to the motionally induced electric current of density* ${}^{str}\mathbf{J}_{str}$ *in the stream measures the physically adequate stimulus the drifting shark receives, whereas the pore-to-pore line integral* $\int -\rho^{str}\mathbf{J}_{str} \cdot \mathbf{ds} + \int -\rho^{str}\mathbf{J}_{anl} \cdot \mathbf{ds} = \approx 1.25 \times \int -\rho^{str}\mathbf{J}_{str} \cdot \mathbf{ds}$ *measures the signal the receptor membranes detect after enhancement of the physically adequate stimulus due to the high resistivity of the animal and the stimulus increasing charges on opposite skin surfaces.*

Note: How the $-\rho^{str}\mathbf{J}_{str} - \rho^{str}\mathbf{J}_{anl}$ field measured by drifting ampullary canals & sensory epithelia relates to the driving motionally induced electric field $\mathbf{v} \times \mathbf{B}_v$ depends on the degree of electrical loading of the stream. The less the charge accumulations at the boundaries of the stream and, hence, the greater the charge accumulations on the skin of the animal are, the higher the $-\rho^{str}\mathbf{J}_{anl}$ is.

~ Shark Drifting, in Stream, Interaction with $\mathbf{B}_h$, (b) Electric Field Vertical

Interaction of the stream and the shark with $\mathbf{B}_h$ leads to a vertically directed $\mathbf{v} \times \mathbf{B}_h$. The mathematic analysis is analogous to the above under interaction with $\mathbf{B}_v$.

$$\text{!!! Ergo vertically}: \int^{mem} \nabla\varphi_{mem} \cdot \mathbf{ds} = \approx 1.25 \times \int -\rho^{str}\mathbf{J}_{str} \cdot \mathbf{ds} \quad \downarrow \text{in negative direction}$$

Vertical ampullary pairs thus measure, by interaction with the horizontal component of the earth's magnetic field, a vertical ohmic field $-\rho^{str}\mathbf{J}_{str} - \rho^{str}\mathbf{J}_{anl}$, which is

dorsoventrally ↓ directed and depends on the magnetic heading of the stream. It is zero when the stream flows due north or south and strongest when the stream flows due east or west.

The great difference between interaction with $\mathbf{B}_h$ and interaction with $\mathbf{B}_v$ is that, due to the seawater-air interface, surface streams are vertically hardly loaded over most of their widths, hence the vertical $-\rho^{str}\mathbf{J}_{str}$ tends to be weak, except near the lateral boundaries of the stream. Moreover, the vertical $-\rho^{str}\mathbf{J}_{str}$ is the *stronger* the closer the stream flows near the magnetic equator.

The electrical charges accumulating at the seawater surface will tend to leak off sideways driven by the electrostatic field of those charges, being higher in the middle of the stream than at the lateral boundaries of the stream.

~ Electrodes Moored in Stream, Interaction with $\mathbf{B}_v$, (a) Electric Field Horizontal

In stream:

$$\mathbf{v}\times\mathbf{B}_v - \rho^{str}\mathbf{J}_{str} + {}^{str}E_{str} = 0 \qquad \text{forces on unit}^{+}\text{ions}$$

$$\mathbf{v}\times\mathbf{B}_v - \rho^{str}\mathbf{J}_{str} = {}^{str}\nabla\varphi_{str} \qquad \text{potential gradients}$$

Salt bridges pore-to-pore line integral along paths through flowing seawater:

$$\text{Path 1:}\quad V_{p2} - V_{p1} = \int\left(\mathbf{v}\times\mathbf{B}_v - \rho^{str}\mathbf{J}_{str}\right)\cdot \mathbf{ds} = \int^{str}\nabla\varphi_{str}\cdot\mathbf{ds}$$

In electrode leads and in sensing elements, respectively: *no electric current!*

$${}^{cnl}\mathrm{E}_{str} + {}^{cnl}\mathrm{E}_{mem} = 0 \text{ and } -{}^{mem}\mathbf{P}_{mem}/\varepsilon\chi + {}^{mem}\mathrm{E}_{mem} = 0 \qquad \text{forces on unit}^{+}\text{ions}$$

$${}^{cnl}\nabla\varphi_{str} + {}^{cnl}\nabla\varphi_{mem} = 0 \text{ and } -{}^{mem}\mathbf{P}_{mem}/\varepsilon\chi = {}^{mem}\nabla\varphi_{mem} \qquad \text{potential gradients}$$

Salt bridges pore-to-pore line integral along paths through electrode leads and sensing elements:

$$\text{Path 2:}\quad V_{p2} - V_{p1} = -{}^{mem}\mathbf{P}_{mem}/\varepsilon\chi = \int\left({}^{cnl}\nabla\varphi_{str} + {}^{cnl}\nabla\varphi_{mem} + {}^{mem}\nabla\varphi_{mem}\right)\cdot\mathbf{ds}$$

Or, since practically $\int -\mathbf{P}/\varepsilon\chi\cdot\mathbf{ds} = \int^{mem}\nabla\varphi_{mem}\cdot\mathbf{ds}$ and $\int({}^{cnl}\nabla\varphi_{mem} + {}^{mem}\nabla\varphi_{mem})\cdot\mathbf{ds} = 0$

$$\text{Path 2:}\quad V_{p2} - V_{p1} = \int^{mem}\nabla\varphi_{mem}\cdot\mathbf{ds} = \int^{cnl}\nabla\varphi_{str}\cdot\mathbf{ds}$$

From path 2 and path 1 equations:

$$\int^{mem}\nabla\varphi_{mem}\cdot\mathbf{ds} = \int^{cnl}\nabla\varphi_{str}\cdot\mathbf{ds} = \int^{str}\nabla\varphi_{str}\cdot\mathbf{ds}$$

$$\text{!!! Ergo horizontally}: \int^{\text{mem}} \nabla\varphi_{\text{mem}} \cdot \mathbf{ds} = \int^{\text{str}} \nabla\varphi_{\text{str}} \cdot \mathbf{ds} \quad \rightarrow \text{in positive direction}$$

In brief: the pore-to-pore path of integration through the salt bridges & sensing elements and the pore-to-pore path through the seawater have the $\nabla\varphi_{\text{str}}$, due to the charges at the lateral boundaries of the stream, in common, leaving the line integrals of ${}^{\text{mem}}\nabla\varphi_{\text{mem}} = -\rho^{\text{str}}\mathbf{J}_{\text{str}} + \mathbf{v} \times \mathbf{B}_{\text{v}} = {}^{\text{str}}\nabla\varphi_{\text{str}}$ for the stationary electrodes and voltmeter to detect.

Horizontally oriented moored electrodes thus measure—by interaction with the vertical component of the earth's magnetic field—in the stream a motional electric field ${}^{\text{str}}\nabla\varphi_{\text{str}}$ due to the electrical charges accumulated at the *lateral boundaries* of the stream by the motionally induced electric current of density ${}^{\text{str}}\mathbf{J}_{\text{str}}$. The ${}^{\text{str}}\nabla\varphi_{\text{str}}$ is normal to the upstream direction → and does not depend on the magnetic heading of the stream.

The horizontal ${}^{\text{str}}\nabla\varphi_{\text{str}}$ is the stronger, the less the stream is electrically loaded, that is, the larger the charge accumulations at the boundaries of the stream are, hence the less the $-\rho^{\text{str}}\mathbf{J}_{\text{str}}$ is, and the more ${}^{\text{str}}\nabla\varphi_{\text{str}}$ counteracts $\mathbf{v} \times \mathbf{B}_{\text{v}}$. Conversely, if the stream is virtually shorted electrically, the ${}^{\text{str}}\nabla\varphi_{\text{str}}$ is nearly nil and the electrodes measure nearly naught.

~ Electrodes Moored in Stream, Interaction with $\mathbf{B}_{\text{h}}$, (b) Electric Field Vertical

Interaction of the stream with $\mathbf{B}_{\text{h}}$ leads to a vertical $\mathbf{v} \times \mathbf{B}_{\text{h}}$. The results are analogous with those under interaction with $\mathbf{B}_{\text{v}}$.

$$\text{!!! Ergo vertically}: \int^{\text{mem}} \nabla\varphi_{\text{mem}} \cdot \mathbf{ds} = \int^{\text{str}} \nabla\varphi_{\text{str}} \cdot \mathbf{ds} \quad \uparrow \text{ or } \downarrow \text{ direction}$$

Vertically oriented moored electrodes thus measure—by interaction with the horizontal component of the earth's magnetic field—in the stream a motional electric field ${}^{\text{str}}\nabla\varphi_{\text{str}}$ due to the electrical charges accumulated at the *surface* and *bottom* of the stream by the motionally induced electric current of density ${}^{\text{str}}\mathbf{J}_{\text{str}}$. The ${}^{\text{str}}\nabla\varphi_{\text{str}}$ is maximally positive ↑ when the stream flows due east, maximally negative ↓ when the stream flows due west, and zero when the stream flows either due north or south.

The great difference between interaction with $\mathbf{B}_{\text{v}}$ and interaction with $\mathbf{B}_{\text{h}}$ is that, due to the seawater-air interface, wind-driven surface streams are vertically hardly loaded over most of their widths, so that the vertically directed ${}^{\text{str}}\nabla\varphi_{\text{str}}$ tends to be strong, hence the $-\rho^{\text{str}}\mathbf{J}_{\text{str}}$ weak, except near the lateral boundaries of the stream.

~ Shark Stationary in Stream, Interaction with $\mathbf{B}_{\text{v}}$, (a) Electric Field Horizontal

The sensory system enabling *stationary* sharks to detect the motional electric fields of ocean streams—consisting of sensory epithelia, conductive gel-filled ampullary canals, and skin pores—and the oceanographic system to measure the electric fields of ocean streams—consisting of *stationary* voltmeters, electrodes, and open-ended agar-filled salt bridge tubes—are very similar. Yet, there is an important difference: in marine sharks the sensory system is embedded in a body of higher resistivity than the seawater, whereas the oceanographic measurement system is directly exposed to the seawater. The skin and body tissues of the sharks make the analysis much more complicated.

Understand well and keep in mind, by introducing the more resistive shark into the less resistive seawater of the stream, the motionally induced electric current of density ${}^{str}\mathbf{J}_{str}$ flowing across the stream causes + and – electrical charges to accumulate on opposite sides of the animal. They are the charges that force part of the originally undisturbed electric current in the stream to flow around the more than seawater resistive animal instead of flowing through it.

We shall first determine (1) how the ampullary system senses the field in the stationary shark, then (2) how the field in the stationary shark relates to the undisturbed field in the ocean stream, and (3) how the signals the stationary shark receives differ from the signals the moored electrodes detect? The following analysis is tedious. Yet, the results are surprisingly simple, almost self-evident: *Stationary sharks detect the voltage difference resulting from the electrical charges accumulated at the boundaries of the stream—like stationary electrodes do—but diminished (!) by the voltage difference between opposite skin surfaces due to the shark being more resistive than the seawater.*

(i) Inside the stationary shark, but outside ampullary sense organs:

$$-\rho^{anl}\mathbf{J}_{str} - \rho^{anl}\mathbf{J}_{anl} + {}^{anl}E_{str} + {}^{anl}E_{anl} = 0 \qquad \text{forces on unit}^{+}\text{ions}$$

$$-\rho^{anl}\mathbf{J}_{str} - \rho^{anl}\mathbf{J}_{anl} = {}^{anl}\nabla\varphi_{str} + {}^{anl}\nabla\varphi_{anl} \qquad \text{potential gradients}$$

The ${}^{anl}\mathbf{E}_{str} = -{}^{anl}\nabla\varphi_{str} \leftarrow$ pertains to the motional electric field in the animal due to the – and + charges at the lateral boundaries of the stream. The ${}^{anl}\mathbf{E}_{str}$ opposes the $\mathbf{v} \times \mathbf{B}_v$. The ${}^{anl}\mathbf{E}_{anl} = -{}^{anl}\nabla\varphi_{anl} \rightarrow$ pertains to the motional electric field in the animal due to the + and – charges on opposite skin surfaces of the animal. The ${}^{anl}\mathbf{E}_{anl}$ aids the $\mathbf{v} \times \mathbf{B}_v$. The charges on opposite sides of the animal are in the stationary shark not only due to the charges diverting the electric current → of the stream partially around the animal, as the shark is more resistive than the seawater medium, but also due to the stationary animal to function as a return path ← for the electric current in the moving stream.

The $-\rho^{anl}\mathbf{J}_{str} \leftarrow$ is the resistive field in the animal due to the electric current of density ${}^{anl}\mathbf{J}_{str} \rightarrow$ resulting from the $\mathbf{v} \times \mathbf{B}_v$ in the stream and the electrical charges at the lateral boundaries of the stream, not taking the charges on opposite sides of the animal into account, with ρ the resistivity of the animal. The $-\rho^{anl}\mathbf{J}_{anl} \rightarrow$ is the resistive field in the animal due to the electric current of density ${}^{anl}\mathbf{J}_{anl} \leftarrow$ resulting from the $\mathbf{v} \times \mathbf{B}_v$ in the animal and the electrical charges on its opposite sides of the animal, not taking the charges on the lateral boundaries of the stream into account, with ρ the resistivity of the animal. Note that ${}^{anl}\mathbf{J}_{anl} \leftarrow$ *subtracts* from the ${}^{anl}\mathbf{J}_{str} \rightarrow$, as it represents the part of the current that does not flow through but around the animal. For the effect of a resistive sphere on a uniform electric current field, refer to the axiom introduced in Appendix 6.

Pore-to-pore line integral along paths through skin & body tissues:

$$\text{Path 1:}\quad V_{p2} - V_{p1} = \int\left(-\rho^{anl}\mathbf{J}_{str} - \rho^{anl}\mathbf{J}_{anl}\right)\cdot \mathbf{ds} = \int\left({}^{anl}\nabla\varphi_{str} + {}^{anl}\nabla\varphi_{anl}\right)\cdot \mathbf{ds}$$

In ampullary canal and in receptor-cell membranes, respectively: *no electric current*!

$${}^{cnl}E_{str} + {}^{cnl}E_{anl} + {}^{cnl}E_{mem} = 0 \text{ and } -{}^{mem}\mathbf{P}_{mem}/\varepsilon\chi + {}^{mem}E_{mem} = 0 \quad \text{forces on unit}^{+}\text{ions}$$

$${}^{cnl}\nabla\varphi_{str} + {}^{cnl}\nabla\varphi_{anl} + {}^{cnl}\nabla\varphi_{mem} = 0 \text{ and } -{}^{mem}\mathbf{P}_{mem}/\varepsilon\chi = {}^{mem}\nabla\varphi_{mem} \quad \text{potential gradients}$$

The ${}^{cnl}\mathbf{E}_{str} = -{}^{cnl}\nabla\varphi_{str}$ ← pertains to the motional electric field in the ampullary canal due to the – and + charges at the lateral boundaries of the stream. The ${}^{cnl}\mathbf{E}_{anl} = {}^{cnl}\nabla\varphi_{anl}$ → pertains to the motional electric field in the ampullary canal due to the + and – charges on opposite skin surfaces of the animal. The ${}^{cnl}\mathbf{E}_{mem} = -{}^{cnl}\nabla\varphi_{mem}$ ← pertains to the motional electric field in the ampullary canal due to the + and – charges on opposite surfaces of the sensory epithelium. The ${}^{mem}\mathbf{E}_{mem} = -{}^{mem}\nabla\varphi_{mem}$ → pertains to the motional electric field in the sensory epithelium due to the + and – charges on opposite skin surfaces of the sensory epithelium.

Pore-to-pore line integral along paths through sense organs:

$$\text{Path 2:}\quad V_{p2} - V_{p1} = \int -{}^{mem}\mathbf{P}_{mem}/\varepsilon\chi \cdot \mathbf{ds}$$
$$= {}^{cnl}\nabla\varphi_{str} + {}^{cnl}\nabla\varphi_{anl} + {}^{cnl}\nabla\varphi_{mem} + {}^{mem}\nabla\varphi_{mem}\big)\cdot \mathbf{ds}$$

Or, since practically $\int -{}^{mem}\mathbf{P}_{mem}/\varepsilon\chi \cdot \mathbf{ds} = \int {}^{mem}\nabla\varphi_{mem} \cdot \mathbf{ds}$ and $\int({}^{cnl}\nabla\varphi_{mem} + {}^{mem}\nabla\varphi_{mem}) \cdot \mathbf{ds} = 0$

$$\text{Path 2:}\quad V_{p2} - V_{p1} = \int {}^{mem}\nabla\varphi_{mem}\cdot \mathbf{ds} = \int\left({}^{cnl}\nabla\varphi_{str} + {}^{cnl}\nabla\varphi_{anl}\right)\cdot \mathbf{ds}$$

Comparing path 2 equation with path 1 equation:

$$\int {}^{mem}\nabla\varphi_{mem}\cdot \mathbf{ds} = \int\left({}^{anl}\nabla\varphi_{str} + {}^{anl}\nabla\varphi_{anl}\right)\cdot \mathbf{ds}$$

$$\text{!!! Ergo horizontally:} \int {}^{mem}\nabla\varphi_{mem}\cdot \mathbf{ds}$$
$$= \int\left({}^{anl}\nabla\varphi_{str} + {}^{anl}\nabla\varphi_{anl}\right)\cdot \mathbf{ds} \quad \rightarrow \text{in positive direction}$$

In brief: the pore-to-pore path of integration through the ampullary canals & receptor cell membranes and the path of integration through the skin & body tissues outside ampullary sense organs have the line integral of ${}^{anl}\nabla\varphi_{str} + {}^{anl}\nabla\varphi_{anl}$, due to the charges at the lateral boundaries of the stream and to the charges accumulating on opposite sides of the sharks body, in common, leaving the line integrals of ${}^{mem}\nabla\varphi_{mem} = -\rho^{anl}\mathbf{J}_{str} - \rho^{anl}\mathbf{J}_{anl}$, which equals the line integral of ${}^{anl}\nabla\varphi_{str} + {}^{anl}\nabla\varphi_{anl}$, for the ampullary sense organs of the stationary animal to detect.

Note that the pore-to-pore line integrals of ${}^{str}\nabla\varphi_{str}$ and ${}^{str}\nabla\varphi_{anl}$ are of opposite sign, + and – respectively. The positive ${}^{str}\nabla\varphi_{str}$ tends to override the negative ${}^{str}\nabla\varphi_{anl}$ when the stream is lightly loaded, that is, when the charge accumulations at the boundaries of the stream are large, hence ${}^{str}\mathbf{J}_{str}$ is small. Conversely, ${}^{str}\nabla\varphi_{anl}$ tends to override the ${}^{str}\nabla\varphi_{str}$ when the stream is heavily loaded, that is, the charge accumulations at the boundaries of the stream are small, hence ${}^{str}\mathbf{J}_{str}$ is large. Hence, in the heavily loaded stream, the ${}^{str}\nabla\varphi_{anl}$ may even reverse the total line integral of ${}^{str}\nabla\varphi_{str} + {}^{str}\nabla\varphi_{anl}$ the animal receives. The ${}^{str}\nabla\varphi_{anl}$ is also the stronger, the more resistive the animal is than the seawater.

(ii) This brings us to the Holy Grail question: how does the ${}^{anl}\nabla\varphi_{str} + {}^{anl}\nabla\varphi_{anl}$ the shark detects relate to the ${}^{str}\nabla\varphi_{str}$ in the stream? The difference from the situation of the shark *drifting* in the ocean stream is that the shark now remains *stationary*, thus offering an additional, local return path for the electric current across the flowing ocean stream. This causes – and + charge accumulations on the left and right side of the animal and an electric current from right to left closing its path in dipole fashion through the seawater of the stream. Since the new charge accumulations are of opposite sign compared to the + and – charge accumulations diverting the current around the animal, the new charge accumulations reduce the total charge accumulations.

In stream:

$$\mathbf{v} \times \mathbf{B}_v - \rho^{str}\mathbf{J}_{str} - \rho^{str}\mathbf{J}_{anl} + {}^{str}E_{str} + {}^{str}E_{anl} = 0 \qquad \text{forces on unit}^{+}\text{ions}$$

$$\mathbf{v} \times \mathbf{B}_v - \rho^{str}\mathbf{J}_{str} - \rho^{str}\mathbf{J}_{anl} = {}^{str}\nabla\varphi_{str} + {}^{str}\nabla\varphi_{anl} \qquad \text{potential gradients}$$

The ${}^{str}\mathbf{E}_{str} = -{}^{str}\nabla\varphi_{str} \leftarrow$ pertains to the motional electric field in the stream due to the – and + charges at the lateral boundaries of the stream. The ${}^{str}\mathbf{E}_{str}$ opposes the $\mathbf{v} \times \mathbf{B}_v$. The ${}^{str}\mathbf{E}_{anl} = -{}^{str}\nabla\varphi_{anl} \rightarrow$ pertains to the motional electric field in the stream due to the + and – charges on opposite skin surfaces of the animal. The ${}^{anl}\mathbf{E}_{anl}$ aids the $\mathbf{v} \times \mathbf{B}_v$.

The $-\rho^{str}\mathbf{J}_{str} \leftarrow$ is the resistive field in the stream due to the electric current of density ${}^{str}\mathbf{J}_{str} \rightarrow$ resulting from the $\mathbf{v} \times \mathbf{B}_v$ in the stream and the electrical charges at the lateral boundaries of the stream, not taking the charges on opposite sides of the animal into account, with ρ the resistivity of the animal. The $-\rho^{str}\mathbf{J}_{anl} \rightarrow$ is the resistive field in the animal due to the electric current of density ${}^{str}\mathbf{J}_{anl} \leftarrow$ resulting from the $\mathbf{v} \times \mathbf{B}_v$ in the animal and the electrical charges on its opposite sides of the animal, not taking the charges on the lateral boundaries of the stream into account, with ρ the resistivity of the animal. Note that ${}^{str}\mathbf{J}_{anl} \rightarrow$*adds* to the ${}^{str}\mathbf{J}_{str} \rightarrow$, as it represents the part of the current that does not flow through, but around the animal. For the effect of a resistive sphere on a uniform electric current field, refer to the axiom introduced in Appendix 6.

Pore-to-pore line integral along paths through flowing seawater:

$$\text{Path 3:}\quad V_{p2} - V_{p1} = \int \left(\mathbf{v} \times \mathbf{B}_v - \rho^{str} \mathbf{J}_{str} - \rho^{str} \mathbf{J}_{anl} \right) \cdot \mathbf{ds} = \int \left({}^{str}\nabla \varphi_{str} + {}^{str}\nabla \varphi_{anl} \right) \cdot \mathbf{ds}$$

In animal:

$$-\rho^{anl}\mathbf{J}_{str} - \rho^{anl}\mathbf{J}_{anl} + {}^{anl}E_{str} + {}^{anl}E_{anl} = 0 \qquad \text{forces on unit}^{+}\text{ions}$$

$$-\rho^{anl}\mathbf{J}_{str} - \rho^{anl}\mathbf{J}_{anl} = {}^{anl}\nabla \varphi_{str} + {}^{anl}\nabla \varphi_{anl} \qquad \text{potential gradients}$$

Pore-to-pore line integral along paths through skin & body tissues:

$$\text{Path 1:}\quad V_{p2} - V_{p1} = \int \left(-\rho^{anl}\mathbf{J}_{str} - \rho^{anl}\mathbf{J}_{anl} \right) \cdot \mathbf{ds} = \int \left({}^{anl}\nabla \varphi_{str} + {}^{anl}\nabla \varphi_{anl} \right) \cdot \mathbf{ds}$$

Comparing path 3 equation with path 1 equation: $-\rho^{anl}\mathbf{J}_{str} - \rho^{anl}\mathbf{J}_{anl} = \mathbf{v} \times \mathbf{B}_v - \rho^{str}\mathbf{J}_{str} - \rho^{str}\mathbf{J}_{rta}$ and ${}^{anl}\nabla\varphi_{str}+{}^{anl}\nabla\varphi_{anl} = {}^{str}\nabla\varphi_{str}+{}^{str}\nabla\varphi_{anl}$

$$\text{!!! Ergo horizontally:} \int {}^{mem}\nabla \varphi_{mem} \cdot \mathbf{ds} = \int \left({}^{str}\nabla \varphi_{str} + {}^{str}\nabla \varphi_{anl} \right) \cdot \mathbf{ds}$$

In brief: whereas expressed in terms of internal body tissue fields, the sense organs of the stationary shark detect the pore-to-pore line integrals of ${}^{mem}\nabla\varphi_{mem}={}^{anl}\nabla\varphi_{str}+{}^{anl}\nabla\varphi_{anl}$, or expressed in terms of the external seawater fields, the sense organs in the stationary shark detect the pore-to-pore line integrals of ${}^{mem}\nabla\varphi_{mem}={}^{str}\nabla\varphi_{str}+{}^{str}\nabla\varphi_{anl}$.

A shark stationary in the stream—by swimming upstream as fast as it drifts downstream—thus receives—by interaction with the vertical component of the earth's magnetic field—a horizontal motional electric field ${}^{str}\nabla\varphi_{str}$ of the charges at the lateral boundaries of the stream *plus* a horizontal electric field ${}^{str}\nabla\varphi_{anl}$ due to the charges accumulating on opposite skin surfaces resulting from the horizontal ${}^{str}\mathbf{J}_{str}$ of the stream. The horizontal ${}^{str}\nabla\varphi_{str}+{}^{str}\nabla\varphi_{anl}$ field is independent of the stream's magnetic heading.

Note that the pore-to-pore line integrals of ${}^{str}\nabla\varphi_{str}$ and ${}^{str}\nabla\varphi_{anl}$ are of opposite sign, + and – respectively. The positive ${}^{str}\nabla\varphi_{str}$ tends to override the negative ${}^{str}\nabla\varphi_{anl}$ when the stream is lightly loaded, that is, when the charge accumulations at the boundaries of the stream are large, hence ${}^{str}\mathbf{J}_{str}$ is small. Conversely, ${}^{str}\nabla\varphi_{anl}$ tends to override the ${}^{str}\nabla\varphi_{str}$ when the stream is heavily loaded, that is, the charge accumulations at the boundaries of the stream are small, hence ${}^{str}\mathbf{J}_{str}$ is large. Hence, in the heavily loaded stream, the ${}^{str}\nabla\varphi_{anl}$ may even reverse the total line integral of ${}^{str}\nabla\varphi_{str}$ $+{}^{str}\nabla\varphi_{anl}$ the animal receives. The ${}^{str}\nabla\varphi_{anl}$ is also the stronger, the more resistive the animal is than the seawater.

(iii) By how much does ${}^{str}\nabla\varphi_{str}+{}^{str}\nabla\varphi_{anl}$ differs from ${}^{str}\nabla\varphi_{str}$ the stationary electrodes detect depends not only on the resistivity of the seawater stream, the average resistivity of the animal's skin and body tissues, and the three-dimensional shape of the animal, but also depends on the loading of the stream, since the

${}^{str}\nabla\varphi_{anl}$ also depends on the ${}^{str}\mathbf{J}_{str}$, which is small when ${}^{str}\nabla\varphi_{str}$ is large and vice versa.

In short, the entire set of charge accumulations effecting the shark are those at the lateral boundaries of the stream, those at the left and right side of the animal due to the electrical current of the stream causing some diminution of the stimulus, and those at the right and left side of the animal due to the loading of the stream by the animal causing a small decrease in diminution.

! ! ! Physically Adequate Stimulus

We thus arrive at the conclusion that, for a shark stationary in an ocean stream, the horizontal $\int {}^{mem}\nabla\varphi_{mem} \cdot \mathbf{ds}$ equals $\int ({}^{str}\nabla\varphi_{str}+{}^{str}\nabla\varphi_{anl}) \cdot \mathbf{ds}$. Hence, *the pore-to-pore line integral* $\int {}^{str}\nabla\varphi_{str} \cdot \mathbf{ds}$ *of the potential gradient due to the charge accumulations at the boundaries of the stream measures the physically adequate stimulus the stationary shark receives, whereas the pore-to-pore line integral* $\int {}^{str}\nabla\varphi_{str} \cdot \mathbf{ds} + \int {}^{str}\nabla\varphi_{anl} \cdot \mathbf{ds}$ *measures the signal the receptor membranes detect after diminution of the physically adequate stimulus due to the stimulus diminishing charges on opposite skin surfaces.*

Note: How the ${}^{str}\nabla\varphi_{str}+{}^{str}\nabla\varphi_{anl}$ field measured by stationary ampullary canals & sensory epithelia relates to the driving motionally induced electric field $\mathbf{v} \times \mathbf{B}_v$ in the stream depends on the electrical loading of the stream. The more the charge accumulations at the boundaries of the stream are and, hence, the less the charge accumulations on the skin of the animal are, the less the ${}^{str}\nabla\varphi_{anl}$ is.

Shark Stationary in Stream, Interaction with $\mathbf{B}_h$, (b) Electric Field Vertical

Interaction of the stream with $\mathbf{B}_h$ leads to a vertically directed $\mathbf{v} \times \mathbf{B}_h$, which depends on the magnetic compass heading of the stream. $\mathbf{v} \times \mathbf{B}_h$ is minimal when the stream flows due north and maximum when the stream flows due east. The results are analogous to those under interaction with $\mathbf{B}_h$ with the main difference that, due to the seawater-air interface, surface streams are hardly loaded over most of their widths, so that ${}^{str}\nabla\varphi_{str}$ tends to be strong, but ${}^{str}\nabla\varphi_{anl}$ tends to be weak, except near the lateral boundaries of the stream.

!!! Ergo vertically: $\int^{mem} \nabla\varphi_{mem} \cdot \mathbf{ds} = \int \left({}^{str}\nabla\varphi_{str} + {}^{str}\nabla\varphi_{anl} \right) \cdot \mathbf{ds} \ \uparrow$ in positive direction

\# A shark stationary in the stream—by swimming upstream as fast as it drifts downstream—thus receives—by interaction with the horizontal component of the earth's magnetic field—a vertical motional electric field ${}^{str}\nabla\varphi_{str}$ of the charges at the top and bottom boundaries of the stream *plus* the vertical electric field ${}^{str}\nabla\varphi_{anl}$ due to the charges accumulating on the ventral and dorsal surfaces of the skin resulting from the vertical ${}^{str}\mathbf{J}_{str}$ of the stream. The vertical ${}^{str}\nabla\varphi_{str}+{}^{str}\nabla\varphi_{anl}$ field depends on the stream's magnetic heading.

Note that the pore-to-pore line integrals of ${}^{str}\nabla\varphi_{str}$ and ${}^{str}\nabla\varphi_{anl}$ are of opposite sign., + and – respectively. The positive ${}^{str}\nabla\varphi_{str}$ tends to override the negative ${}^{str}\nabla\varphi_{anl}$ as the stream is vertically lightly loaded, that is, the charge accumulations at the top and bottom of the stream are large, hence ${}^{str}\mathbf{J}_{str}$ is small. The ${}^{str}\nabla\varphi_{anl}$ is the stronger, the more resistive the animal is than the seawater.

~ Electrodes Moored on Bottom, Under Stream

Under stream:

$$-\rho^{stw}\mathbf{J}_{str} + {}^{stw}\mathrm{E}_{str} = 0 \qquad \text{forces on unit}^{+}\text{ions}$$

$$-\rho^{stw}\mathbf{J}_{str} = {}^{stw}\nabla\varphi_{str} \qquad \text{potential gradients}$$

Salt bridges pore-to-pore line integral along paths through still seawater:

$$\text{Path 1:}\quad V_{p2} - V_{p1} = \int -\rho^{stw}\mathbf{J}_{str}\cdot\mathbf{ds} = \int^{stw}\nabla\varphi_{str}\cdot\mathbf{ds}$$

In electrode leads and in sensing elements, respectively: *no electric current*!

$${}^{cnl}\mathrm{E}_{str} + {}^{cnl}\mathrm{E}_{mem} = 0 \text{ and } -{}^{mem}\mathbf{P}_{mem}/\varepsilon\chi + {}^{mem}\mathrm{E}_{mem} = 0 \qquad \text{forces on unit}^{+}\text{ions}$$

$${}^{cnl}\nabla\varphi_{str} + {}^{cnl}\nabla\varphi_{mem} \text{ and } -{}^{mem}\mathbf{P}_{mem}/\varepsilon\chi = {}^{mem}\nabla\varphi_{mem} \qquad \text{potential gradients}$$

Salt bridges pore-to-pore line integral along paths through electrode leads and sensing elements:

$$\text{Path 2:}\quad V_{p2} - V_{p1} = \int -{}^{mem}\mathbf{P}_{mem}/\varepsilon\chi\cdot\mathbf{ds} = \int\left({}^{cnl}\nabla\varphi_{str} + {}^{cnl}\nabla\varphi_{mem} + {}^{mem}\nabla\varphi_{mem}\right)\cdot\mathbf{ds}$$

Or, since practically $\int -{}^{mem}\mathbf{P}_{mem}/\varepsilon\chi\cdot\mathbf{ds} = \int^{mem}\nabla\varphi_{mem}\cdot\mathbf{ds}$ and $\int({}^{cnl}\nabla\varphi_{mem} + {}^{mem}\nabla\varphi_{mem})\cdot\mathbf{ds} = 0$

$$\text{Path 2:}\quad V_{p2} - V_{p1} = \int^{mem}\nabla\varphi_{mem}\cdot\mathbf{ds} = \int^{cnl}\nabla\varphi_{str}\cdot\mathbf{ds}$$

From path 1 and path 2 equations:

$$\text{!!! Ergo:}\int^{mem}\nabla\varphi_{mem}\cdot\mathbf{ds} = \int^{stw}\nabla\varphi_{str}\cdot\mathbf{ds} \qquad \rightarrow \text{in positive direction}$$

In brief: the pore-to-pore path of integration through salt bridges & sensing elements and the pore-to-pore path through the seawater have $\nabla\varphi_{str}$ in common, leaving the line integrals of $\int^{mem}\nabla\varphi_{mem}\cdot\mathbf{ds} = -\rho^{stw}\mathbf{J}_{str} = \int^{stw}\nabla\varphi_{str}\cdot\mathbf{ds}$ for the voltmeter to detect.

Bottom electrodes thus measure under the stream a horizontal motional electric field ${}^{stw}\nabla\varphi_{str}$ due to the volume charges at the boundaries of the stream. The ${}^{stw}\nabla\varphi_{str}$ is normal to the upstream direction → and does not depend on the magnetic heading of the stream.

Note: If the stream extends to the bottom, then the electrodes are in the stream and the ocean bottom provides the only return path.

Note: If the stream is still, then the electrodes may still measure the electric field of any other, arbitrary charge accumulations.

~ Shark Resting on Bottom, Under Stream

Under the stream:

$$-\rho^{stw}\mathbf{J}_{str} - \rho^{stw}\mathbf{J}_{anl} + {}^{stw}\mathrm{E}_{str} + {}^{stw}\mathrm{E}_{anl} = 0 \qquad \text{forces on unit}^{+}\text{ions}$$

$$-\rho^{stw}\mathbf{J}_{str} - \rho^{stw}\mathbf{J}_{anl} = {}^{stw}\nabla\varphi_{str} + {}^{stw}\nabla\varphi_{anl} \qquad \text{potential gradients}$$

Electrode-to-electrode line integral along paths through seawater:

$$\text{Path 1:} \quad V_{p2} - V_{p1} = \int\left(-\rho^{stw}\mathbf{J}_{str} - \rho^{stw}\mathbf{J}_{anl}\right)\cdot\mathbf{ds} = \int\left({}^{stw}\nabla\varphi_{str} + {}^{stw}\nabla\varphi_{anl}\right)\cdot\mathbf{ds}$$

In ampullary canal and in receptor-cell membrane, respectively: *no electric current*!

$${}^{cnl}E_{str} + {}^{cnl}E_{anl} + {}^{cnl}E_{mem} = 0 \text{ and } -{}^{mem}\mathbf{P}_{mem}/\varepsilon\chi + {}^{mem}E_{mem} = 0 \quad \text{forces on unit}^{+}\text{ions}$$

$${}^{cnl}\nabla\varphi_{str} + {}^{cnl}\nabla\varphi_{anl} + {}^{cnl}\nabla\varphi_{mem} = 0 \text{ and } -{}^{mem}\mathbf{P}_{mem}/\varepsilon\chi = {}^{mem}\nabla\varphi_{mem} \quad \text{potential gradients}$$

Pore-to-pore line integral along paths through the sense organs:

$$\text{Path 2:} \quad V_{p2} - V_{p1} = \int -{}^{mem}\mathbf{P}_{mem}/\varepsilon\chi\cdot\mathbf{ds}$$
$$= \int\left({}^{cnl}\nabla\varphi_{str} + {}^{cnl}\nabla\varphi_{anl} + {}^{cnl}\nabla\varphi_{mem} + {}^{mem}\nabla\varphi_{mem}\right)\cdot\mathbf{ds}$$

Or, since practically $\int -{}^{mem}\mathbf{P}_{mem}/\varepsilon\chi \cdot \mathbf{ds} = \int {}^{mem}\nabla\varphi_{mem} \cdot \mathbf{ds}$ and $\int({}^{cnl}\nabla\varphi_{mem} + {}^{mem}\nabla\varphi_{mem}) \cdot \mathbf{ds} = 0$

$$\text{Path 2:} \quad V_{p2} - V_{p1} = \int {}^{mem}\nabla\varphi_{mem}\cdot\mathbf{ds} = \int\left({}^{stw}\nabla\varphi_{str} + {}^{stw}\nabla\varphi_{anl}\right)\cdot\mathbf{ds}$$

From path 1 and path 2 equations:

$$\text{!!! Ergo:} \int {}^{mem}\nabla\varphi_{mem}\cdot\mathbf{ds} = \int\left({}^{stw}\nabla\varphi_{str} + {}^{stw}\nabla\varphi_{anl}\right)\cdot\mathbf{ds} \qquad \rightarrow \text{in positive direction}$$

In brief: the pore-to-pore path of integration through ampullary canals & receptor cell membranes and the pore-to-pore path through skin and body tissues have the line integrals of $\nabla\varphi_{str}$ and $\nabla\varphi_{anl}$ in common, leaving the line integrals of $^{mem}\nabla\varphi_{mem} = -\rho^{stw}\mathbf{J}_{str} - \rho^{stw}\mathbf{J}_{anl} = {}^{stw}\nabla\varphi_{str} + {}^{stw}\nabla\varphi_{anl}$ for the ampullary sense organs of the shark resting on the bottom to detect.

Sharks resting on the bottom under the stream thus measure the motional electric field $^{stw}\nabla\varphi_{str}$ due to the volume charges at the boundaries of the stream, *plus* the electric field $^{stw}\nabla\varphi_{anl}$ of the surface charges accumulating on the skin due to the return current of the stream $^{stw}\mathbf{J}_{str}$. The $^{stw}\nabla\varphi_{str} + {}^{stw}\nabla\varphi_{anl}$ is normal to the upstream direction → and does not depend on the magnetic heading of the stream.

Paradoxically, under the stream, $\int {}^{stw}\nabla\varphi_{str} \cdot \mathbf{ds} = \int -\rho^{stw}\mathbf{J}_{str} \cdot \mathbf{ds}$ is the larger, the less heavily the stream is electrically loaded, that is, the smaller the total return current is, and the larger the charge accumulations at the boundaries of the stream are.

When the stream is all but shorted, that is, when the stream is extremely shallow compared to the great depth of the ocean, $^{stw}\nabla\varphi_{str}$ tends to 0 and $-\rho^{str}\mathbf{J}_{str}$ tends to $-\mathbf{v} \times \mathbf{B}_v$. Hence, by lack of charge accumulations at the boundaries of the stream, $^{stw}\nabla\varphi_{str} = -\rho^{stw}\mathbf{J}_{str}$ tends to 0, not due to lack of return current, but to lack of return current density $^{stw}\mathbf{J}_{str}$ along the deep return path.

When the stream is all but insulated, that is, when the return path is extremely shallow compared to the great depth of the stream, $^{str}\nabla\varphi_{str}$ tends to $-\mathbf{v} \times \mathbf{B}_v$ and $-\rho^{str}\mathbf{J}_{str}$ tends to 0. Hence, by the large amount of charge accumulations at the boundaries of the stream, $^{stw}\nabla\varphi_{str} = -\rho^{stw}\mathbf{J}_{str}$ is high, not due to the large return current, but to the large current density $-\rho^{stw}\mathbf{J}_{str}$ along the shallow return path.

Note: If the stream extends to the bottom, then the shark is in the stream and the ocean crust provides the only return path.

Orders of Magnitude

To gage—in a shark drifting in an ocean stream—the strength of the $^{mem}\mathbf{E}_{mem}$ acting on the gating charges of the voltage-sensitive ion channels in the receptor membranes, expressed in terms of the physically adequate stimulus $-\rho^{str}\mathbf{J}_{str}$, enhanced nominally 1.25 times due to the charge accumulations on opposite sides of the animal, let the distance between the skin pores of a differential pair of ampullae be 20 cm, and let the two opposing sensory epithelia—for now—consist of just two single membranes with a combined thickness of 10 nm. Hence, at an average $-\rho\mathbf{J}$ threshold field of 5 nV/cm in the body tissues outside the ampullary sense organs, a value proven to be readily detectable by the animal, the voltage difference between the 20 cm apart skin pores measures 100 or 50 nV for each of the two sensory membranes. However, since the sensory membranes are ≈200,000,000 × thinner than the ampullary canals are long, the voltage gradient in the sensory membranes is 5 nV/cm × 200,000,000 = 1 V/cm. For a shark drifting in a heavily loaded ocean stream, the averaged 5 nV/cm in the body tissues corresponds to a $-\rho^{str}\mathbf{J}_{str}$ field of ≈4 nV/cm, which at a vertical component of the earth's magnetic field of 25 μT corresponds to a drift velocity of some 2 cm/s, conservatively rated.

Since the afferent electrophysiological nerve response is approximately linear from the threshold value to eight times the threshold value, the dynamic range of the individual longest canal ampullae is ≈ 5–40 nV/cm. Nevertheless, the dynamic range of the ampullary system is much larger, since the pore-to-pore distances of differential ampullae vary from very small to great lengths. Below the narrow, eightfold dynamic range of the individual ampullary sense organs, the signal is swamped by the system noise. Above this range, the individual ampullae quickly saturate, as judged from the afferent action-potential rates. Amplification of the electrical signal resulting from the negative receptor current causing positive feedback will be discussed under Sect. 2.3.2.

2.1.3 *Further Developments: Physical Features of Ocean Streams*

2.1.3.1 Effect of Loading by Return Current Through Deeper Water, Revisited

In a shallow, electrically heavily loaded open-ocean stream, the charge accumulations at the boundaries of the *stream* are small. Hence, the resulting ${}^{str}\nabla\varphi_{str}$ is weak and the strength of $-\rho^{str}\mathbf{J}_{str}$ is only a little less than the strength of $-\mathbf{v} \times \mathbf{B}_v$. However, due to the sizeable charge accumulations on opposite skin surfaces of the *animal*, resulting from the strong electrical current of density ${}^{str}\mathbf{J}_{str}$, the ${}^{str}\nabla\varphi_{anl}$ of the drifting shark is nearly at its strongest. Hence, the $-\rho^{str}\mathbf{J}_{anl}$ is not at all insignificant. As a result, the pore-to-pore $\int\nabla\varphi_{mem} \cdot \mathbf{ds}$ equaling $\int(-\rho^{str}\mathbf{J}_{str} - \rho^{str}\mathbf{J}_{anl}) \cdot \mathbf{ds}$ may even be stronger than the $-\int \mathbf{v} \times \mathbf{B}_v \cdot \mathbf{ds}$ in a heavily loaded stream. At the other extreme, in a tidal stream in a fjord, the ${}^{str}\nabla\varphi_{str}$ nearly nulls the oppositely directed $\mathbf{v} \times \mathbf{B}_v$ in strength, and, due to the small current density ${}^{str}\mathbf{J}_{str}$, the $-\rho^{str}\mathbf{J}_{str}$ is almost zero, leaving hardly anything for the drifting animal to detect, even though $\int\mathbf{v} \times \mathbf{B}_v \cdot \mathbf{ds}$ is large. This dramatically demonstrates that not the pore-to-pore line integral of $-\mathbf{v} \times \mathbf{B}_v$, but the pore-to-pore line integral of $-\rho^{str}\mathbf{J}_{str}$ represents the physically adequate stimulus, the two being the more different in strength, the less the stream is loaded. The $-\rho^{str}\mathbf{J}_{anl}$ makes what the animal detects even larger than merely the $-\rho^{str}\mathbf{J}_{str}$, given that the animal is more resistive than the surrounding water.

2.1.3.2 Critical Importance of Charge Accumulations, Revisited

The space charges at the lateral boundaries of the stream diminish the $-\rho\mathbf{J}_{str}$, rendering the stream less detectable by the drifting shark, whereas the surface charges on the skin of the animal enhance the $-\rho\mathbf{J}_{anl}$, facilitating the detection of the stream by the drifting shark. To summarize:

Let us consider three levels of electrically loading an ocean stream by the return current through deeper water, first for bare drifting and moored *electrodes* and then for ampullary sense organs embedded in a stationary and in a drifting *shark*.

In the stream, bare *drifting* electrodes measure $\int -\rho\mathbf{J}_{str} \cdot \mathbf{ds}$, which equals $\int(-\mathbf{v} \times \mathbf{B}_v + \nabla\varphi_{str}) \cdot \mathbf{ds}$.

If in the limit the stream is so heavily loaded that, by lack of charges at the boundaries of the stream, $\int\nabla\varphi_{str} \cdot \mathbf{ds}$ is zero, then the $\int -\rho\mathbf{J}_{str} \cdot \mathbf{ds}$ the electrodes measure equals $-\int \mathbf{v} \times \mathbf{B}_v \cdot \mathbf{ds}$. Unrealistic!

If in the limit the stream is not loaded at all, so that by lack of electric current across the stream $\int -\rho\mathbf{J}_{str} \cdot \mathbf{ds}$ the electrodes measure is zero, then $\int\mathbf{v} \times \mathbf{B}_v \cdot \mathbf{ds}$ equals $\int\nabla\varphi_{str} \cdot \mathbf{ds}$. Realistic.

If the stream is moderately loaded so that $\int\mathbf{v} \times \mathbf{B}_v \cdot \mathbf{ds}$ is balanced partially by $\int -\rho\mathbf{J}_{str} \cdot \mathbf{ds}$ and partially adds to $\int\nabla\varphi_{str} \cdot \mathbf{ds}$, then the $\int -\rho\mathbf{J}_{str} \cdot \mathbf{ds}$ the electrodes measure equals a fraction of $-\int \mathbf{v} \times \mathbf{B}_v \cdot \mathbf{ds}$.

In the stream, *moored* electrodes measure $\int\nabla\varphi_{str} \cdot \mathbf{ds}$, which equals $\int(\mathbf{v} \times \mathbf{B}_v - \rho\mathbf{J}_{str}) \cdot \mathbf{ds}$.

If in the limit the stream is so heavily loaded that, by lack of charges at the boundaries of the stream, $\int\nabla\varphi_{str} \cdot \mathbf{ds}$ the electrodes measure is zero, then $\int -\rho\mathbf{J}_{str} \cdot \mathbf{ds}$ equals $-\int \mathbf{v} \times \mathbf{B}_v \cdot \mathbf{ds}$. Unrealistic!

If in the limit the stream is not loaded at all, so that by lack of current of density $\mathbf{J}_{str}$ across the stream $\int -\rho\mathbf{J}_{str} \cdot \mathbf{ds}$ is zero, then the $\int\nabla\varphi_{str} \cdot \mathbf{ds}$ the electrodes measure equals $-\int \mathbf{v} \times \mathbf{B}_v \cdot \mathbf{ds}$. Realistic.

If the stream is moderately loaded, so that $-\int \mathbf{v} \times \mathbf{B}_v \cdot \mathbf{ds}$ is partially balanced by $\int -\rho\mathbf{J}_{str} \cdot \mathbf{ds}$ and partially adds to $\int\nabla\varphi_{str} \cdot \mathbf{ds}$, then the $\int\nabla\varphi_{str} \cdot \mathbf{ds}$ the electrodes measure equals a fraction of $\int\mathbf{v} \times \mathbf{B}_v \cdot \mathbf{ds}$.

A shark drifting in the stream detects, in addition to what drifting electrodes measure, the $\int\nabla\varphi_{anl} \cdot \mathbf{ds}$ of the surface charges on opposite skin surfaces resulting from the $\mathbf{J}_{str}$, when present.

A shark swimming against the stream, thereby *remaining still relative to a moored buoy*, detects, in addition to what moored electrodes measure, the $\int\nabla\varphi_{anl} \cdot \mathbf{ds}$ of the surface charges on opposite skin surfaces resulting from the $\mathbf{J}_{str}$.

2.1.3.3 Effect of Horizontal Component of Earth's Magnetic Field on Ocean Stream, Revisited

Concerning the interaction between the stream and the horizontal component of the earth's magnetic field $\mathbf{B}_h$, the resulting motionally induced electric field $\mathbf{v} \times \mathbf{B}_h$ is vertically directed, the return path for the induced electric current is blocked by the air-water interface, causing surface charges to accumulate at the surface and space charges at the bottom of the stream. Otherwise, the situation is governed by the same physical principles, as the reader may readily ascertain. Hence, in the vertical direction, a shallow, wide ocean stream is hardly loaded, except for the small electric currents flowing from the central region of the stream to the left and right, consisting of the surface charges leaking off horizontally. Hence, far from the lateral boundaries of the stream, the vertical currents are very small, but near the lateral boundaries, where the vertical electric currents can readily return through adjacent still waters, they form significant eddies.

Right at the magnetic equator, where the earth's magnetic field is horizontal, only the eddy currents originating at the edges of the stream occur. Since wind-driven ocean streams are typically on the order of only 100-m-deep but 100-km-wide, the electric fields of the eddies remain over most of the width of the stream extremely weak. The eddies at the left and right side of the stream are opposite in direction and depend in their strength on the compass direction of the stream, being maximal when the stream is heading due east or west. Off the equator, toward northern or southern latitudes, where the horizontal component of the earth's magnetic field gets weaker and the vertical component stronger, the horizontal magnetic component diminishes in significance compared to the effect of the vertical component.

2.1.3.4 Strength of Oceans' Motional Electric Fields

To give an idea of magnitudes, shallow wind-driven ocean streams, such as the well-loaded Gulf Stream, may locally be as fast as 100 cm/s, causing $-\rho\mathbf{J}$ fields on the order of 500 nV/cm measured with towed electrodes, and commonly are not slower than 1 cm/s, causing $-\rho\mathbf{J}$ fields as weak as 5 nV/cm, still enough for sharks and rays to detect. Deep-water streams tend to be more voluminous and to flow slower. Hence, their $-\rho\mathbf{J}$ fields are much weaker than those of surface flows. The periodic tidal streams in the English Channel are fast, but not heavily loaded, as they extend to the bottom, which is more resistive than the seawater. Ocean swells and waves also produce periodic $-\rho\mathbf{J}$ fields, which sharks and rays must be very familiar with. Turbulence causes random fields on a smaller scale and may in part escape the very low-frequency electric sense of sharks and rays. Yet, all are governed by the same mathematical-physics equations.

2.1.4 *Relaxing the Constraints, Extending Models to Meet Nature*

As promised, the simplifying assumptions will be relaxed so as to bridge the gap between our model stream and real ocean streams concerning their interaction with the vertical component of the earth's magnetic field $\mathbf{B}_v$. If the stream is lengthwise uniform and of semi-elliptical cross-section, the resistivity of the water is homogenous, and $\mathbf{v} \times \mathbf{B}_v$ is uniform, then $\mathbf{E}_{str}$ and $-\rho\mathbf{J}_{str}$ are in the stream uniform as well (Fig. 2.1). If the cross-section is of any other form than semi-elliptical, but the $\mathbf{v} \times \mathbf{B}_v$ is still uniform, the $\mathbf{E}_{str}$ and $-\rho\mathbf{J}_{str}$ will get curved, though such as to keep their vector sum uniform in conformity with the $\mathbf{v} \times \mathbf{B}_v$ they are in equilibrium with.

Temporal changes in the speed of the stream cause, by interaction with the vertical component of the earth's magnetic field $\mathbf{B}_v$ proportional changes in the space charge accumulations, divergence that is, at the lateral boundaries of the stream, and proportional changes in the vorticity, curl that is, at the bottom and top of the stream. Temporal changes in the direction of the stream cause the charge accumulations and the vorticity at the boundaries of the stream to readjust such as to turn the plane in which the fields occur over the same angle as the stream turns. If the velocity of the stream does not remain uniform along its length, electrical loop currents occur between upstream and downstream regions of the stream. Interaction with the horizontal component of the earth's magnetic field $\mathbf{B}_h$ is more involved, as it causes vorticity within the stream upon turning left or right, hence circulating loop currents around the velocity axis of the stream and depending on the stream's magnetic compass heading. We will leave those for the sharks to interpret.

Where inside the stream the spatial distribution of the velocities varies, corresponding local variations in the electric currents occur. A hint as to the effects of the changes in velocity may be given as follows. If the velocity $\mathbf{v}$ locally changes in the direction of $\mathbf{v}$, the change in $\mathbf{v} \times \mathbf{B}$ will cause local + and – space charges to accumulate along an axis normal to $\mathbf{v}$. If the velocity $\mathbf{v}$ locally changes in a direction normal to $\mathbf{v}$, the change in $\mathbf{v} \times \mathbf{B}$ will cause local vorticity with its axis parallel with $\mathbf{v}$. The two combined produce local electrical eddy currents within the stream, similar to the grand electric current circuit of the model stream in its entirety. Conceptually, one may anticipate the results, but quantitatively they are hard to obtain. Here again, the animals sure have more experience with these local fields than you and I do.

Non-homogeneous seawater resistivities, temperature differences, and temporal changes in flow velocity are all adding to the complexity of the fields. Yet, they are of secondary relevance to the fundamental principles of the electric and magnetic orientation theory and will here not be discussed further. When sharks are passively moved about by more complicated flows, such as swells, waves, and eddies, including turbulent flows, the electric fields they receive, though still governed by the physical principles outlined thus far, may be more intricate than we can readily envision. Yet, the more complicated the fields are, the greater wealth of information they offer, which remains to be revealed by future scientists. After all, the animals need not follow our intricate reasoning, but unwittingly rely on innate and empirically gained experience.

As for the body of the drifting animal itself, the non-homogeneous resistivity of the skin and body tissues may be thought to complicate the situation dramatically, since the $-\rho\mathbf{J}_{anl}$ depends inherently on the highly nonhomogeneous tissue resistivity ρ. However, when we state that the animal detects the $-\rho\mathbf{J}_{str} - \rho\mathbf{J}_{anl}$, we mean, it detects their pore-to-pore line integrals across the body. And, since at uniform $\mathbf{v} \times \mathbf{B}_v$, the $-\rho\mathbf{J}_{str}$ and $-\rho\mathbf{J}_{anl}$ are conservative, this pore-to-pore line integral is a definite quantity, however complicated $-\rho\mathbf{J}_{str}$ and $-\rho\mathbf{J}_{anl}$ may be inside the animal, see Appendix 4. As one of my keen physical-oceanography colleagues, who struggled with the issue, remarked with great relief: "I see, the animal makes excellent use of the integrative properties of the system." Actually, the pore-to-pore line integral of $-\rho\mathbf{J}_{str} - \rho\mathbf{J}_{anl}$ through the body tissues, with ρ the inhomogeneous resistivity of the tissues, is the same as the pore-to-pore line integral of $-\rho\mathbf{J}_{str} - \rho\mathbf{J}_{anl}$ through the surrounding seawater, with ρ the homogeneous resistivity of the seawater, which may conceptually be a more friendly way of looking at the effective stimulus.

Since the physically adequate stimulus is given by the pore-to-pore line integral of $-\rho\mathbf{J}_{str}$, the skin, by offering much higher resistance than the tissues and body fluids, plays an important part in producing the effective stimulus. For the short ampullae in the snout of sharks, the path through skin is by far the main stretch that generates the $-\rho\mathbf{J}$ stimulus. For the longer canals, the body tissues and fluids play a significant role as well.

One last qualification: the shark was thus far considered an inanimate, resistive object. The electrical signals we have contemplated thus far were in addition to the ongoing electrophysiological phenomena within and at the skin of the animal. It is indeed amazing that, nevertheless, the animal can deal with the astoundingly small electrical sensory signals in the ubiquitous presence of the much stronger electrical phenomena ongoing in the body of the animal itself. The integral properties of the system deserve for that great credit, by paving over local irregularities. For the excitation process and the common mode suppression of the electrical signals, see below.

2.1.5 Biological Interpretation: Orientation in Ocean Streams

2.1.5.1 Drift in Ocean Streams

What may a drifting shark learn from the $\int -\rho\mathbf{J}_{str}\cdot\mathbf{ds}$ it detects? Might it sense the velocity $\mathbf{v}$ of its drift with the stream? But would that not violate the Galilean principle of relativity? No, it does not, because the shark must by experience assume or by exploration determine that the $-\rho\mathbf{J}_{str}$ results from the relative velocity between the stream and deeper water or the ocean bottom, and to what extent it does so. The return current electrically peeks, so to speak, into the deeper water and the ocean bottom with respect to which the stream moves. Still, the shark must take into account by how much $\int -\rho\mathbf{J}_{str}\cdot\mathbf{ds}$ differs from the $-\int \mathbf{v}\times\mathbf{B}_{v}\cdot\mathbf{ds}$, which depends on the amount of space charges at the boundaries of the stream, hence on the degree of loading of the stream. Of course, the animal cannot and need not be consciously aware of the remote space charge accumulations. Yet, it must by exploring the circumstances be able to determine the relationship between the local $-\rho\mathbf{J}_{str}$ and the velocity $\mathbf{v}$ of the stream, which it may as will be shown below. In a heavily loaded open-ocean stream, the lack of charges at the lateral boundaries renders the $-\rho\mathbf{J}_{str}$ maximal, whereas in a hardly loaded rocky fjord, the abundance of charges at the lateral boundaries renders the $-\rho\mathbf{J}_{str}$ minimal.

Strictly speaking, in its own frame of reference, the shark senses the deeper water and the ocean bottom to move in the $-\mathbf{v}$ direction. The shark works more like Google Earth. When it wants to travel from here to there, it simply turns the whole world to draw its destination toward itself. We do the same, but we intelligently conclude that we are moving, and the earth stands still, does it really? Yet, for a shark to detect its drift with the stream implies a transformation between two frames of reference that we cannot assume lightly. Hence, when we say the animal detects its drift, we must be aware that it is our earth-bound view, not necessarily the animal's.

How might a shark, by exploring the depth profile of an open ocean or tidal stream, interpret the $-\rho\mathbf{J}_{str}$ field in terms of the direction and velocity of its drift? For the present discourse, the issue is a matter of physical feasibility, not a matter of reading the animal's mind. We know from the author's behavioral experiments that sharks detect the electric fields of ocean currents, detect the earth's magnetic field, and have the ability to respond to them appropriately. There may be ambiguity and uncertainty, but that does not prevent them from making the best use possible from the limited information available. Actually, no sense organs of any modality provide complete information. To convince ourselves that sharks indeed are able to collect the information they need to derive their drift velocity **v** from the electric fields they receive, let us analyze three idealized situations: a shallow, wind-driven, open-ocean stream, an ocean or tidal stream flowing over shallow waters, and a tidal stream flowing in a non-conductive rocky channel, as in a fjord.

2.1.5.2 Three Situations, Challenging Our Understanding

(i) In the world-wide open oceans, heavily loaded, wind-driven surface streams prevail. In those streams, the $-\rho\mathbf{J}_{str}$ gives the shark a good measure of the upstream direction and the speed of its drift. To verify the situation, the shark may dive down until the ambient $-\rho\mathbf{J}$ field has gone through zero in strength and reversed in direction relative to the direction of the earth's magnetic field, thus giving the animal the depth of the stream. If, when diving deeper and deeper, the bottom is not reached before long, the surface stream was indeed shallow compared to the depth of the ocean, thus heavily loaded. This then assures that in the stream the lessening of the $-\rho\mathbf{J}_{str}$ due to the charge accumulations at the boundaries of the stream was only slight. Hence, that the $-\rho\mathbf{J}_{str}$ signal it receives after returning to the surface stream reasonably well reflects the upstream direction and the strength of the stream, differing from the $-\mathbf{v} \times \mathbf{B}_v$ only by the weak $\mathbf{E}_{str}$ field of the small space charges at the boundaries of the stream. Traveling sharks indeed often make deep dives, cf. Carey et al. (1990).

(ii) However, if the animal reaches an insulating ocean floor soon after diving below the stream, say, after diving as deep under the stream as the stream is deep, then the return path offers roughly as much resistance as the stream itself does. Hence, the stream is then about halfway loaded, that is, half of the $\mathbf{v} \times \mathbf{B}_v$ is counteracted by the $-\rho\mathbf{J}_{str}$ due to the current of density $\mathbf{J}_{str}$ and half by the $\mathbf{E}_{str}$ due to the charge accumulations at the boundaries of the stream. This also means that the return current $\mathbf{J}_{rts}$, driven by the charges at the boundaries of the stream, causes under the stream a $-\rho\mathbf{J}_{rts}$ field about equally strong as the $-\rho\mathbf{J}_{str}$ in the stream, but in the opposite direction. Thus, by taking the vectorial difference between the signal $-\rho\mathbf{J}_{str}$ it received in the stream and the oppositely directed signal $-\rho\mathbf{J}_{rts}$ it received near the bottom, it will gain full information of the direction and velocity of its drift in the surface stream relative to the deeper water. But was the deeper water stationary? If not, the animal will upon landing on the bottom sense a change in the field it receives given by $\mathbf{v} \times \mathbf{B}_v$ and, in addition, the effect of the change in charge accumulations on the skin. After landing, the animal may furthermore fluid-dynamically sense the velocity of the water relative to the bottom, which it then must add to that of the surface stream with respect to deeper water.

(iii) In a bottom-deep, uniform-velocity stream, flowing through a rocky channel, the $-\rho\mathbf{J}_{str}$ is vanishingly small, since the motional electric field $\mathbf{E}_{str}$ of the large charge accumulations on the side walls of the channel all but completely counteracts the $\mathbf{v} \times \mathbf{B}_v$ in the stream. Hence, a drifting shark is not significantly stimulated electrically. However, when the animal dives deep and lands on the bottom, the $\mathbf{v} \times \mathbf{B}_v$ in the animal drops out of the equation and the animal only detects the $\nabla\varphi_{str}$ of the charge accumulations at the boundaries of the stream, which equals the $\mathbf{v} \times \mathbf{B}_v$ field in the stream, besides the effect of the charges accumulating on the skin

Warning: The author does not assert that sharks follow the foregoing procedures. The worked-out examples were only meant as a challenge for the author and the reader to convince themselves that the necessary information is available. Yet, it would be hard to believe that the animals would not "know" how to interpret the information they receive even though it presents biologically vital information. Naturally, the peripheral sense organs merely transduce the weak electrical signals into mere nerve action potentials, whereas the interpretation takes place in the central nervous system, based on information received from all contributing sensory systems, on memory, and experience, learned and innate. Sharks may be smarter than people think they are. I hope their teachings are not wasted on us. I am humiliated by the great effort it takes me to understand these "lower" animals.

2.1.5.3 Remarks

Actually, one other situation must be taken into account. In equatorial waters, the vertical component $\mathbf{B}_v$ of the earth's magnetic field tends to zero, so that the horizontal component $\mathbf{B}_h$ becomes most prominent. Then, instead of the electric fields in the main body of the stream, those near its lateral boundaries prevail. Interestingly, those fields not only indicate the fringes of the streams, but also depend on the magnetic compass direction of the stream. This means, even more complications for us or, in other words, even more information for animals that evolved living in a rich electrical world, in general, non-physical oceanographers are sensibly unfamiliar with.

Naturally, there are many other kinds of ambient electric fields that sharks may detect, such as the fringe fields of the Gulfstream return currents invading the stationary waters as in Bimini Bay, where the author studied the orientation of sharks and rays to ambient electric fields, furthermore the solar-induced magnetically driven electric currents along the rims of large oceans, the fields of tidal waves and upwellings. For an early introduction to the various fields the animals detect, including ubiquitous electrochemical fields, and what the animals may learn from them, see the author's 1974 handbook chapter. For the animals, though, it is of no concern where the environmental fields originate from; they cannot and need not know. They merely have learned to survive and thrive by taking advantage of the information the various electric fields present. The discourse thus far was primarily meant to elucidate the physical principles of the biologically relevant information offered by the fields of wind-driven ocean streams. They were discussed not only because of their intrinsic biological significance, but also as an introduction to the issue of sharks sensing their magnetic compass headings by orienting to the electric fields they induce by moving relative to the water in the presence of the earth's magnetic field, to be discussed in full detail below.

2.1.6 Ocean Streams Observed in Frame of Drifting Shark

In the introduction, I promised first to observe a shark drifting from the arm-chair, earth-bound point of view and then from the point of view of the animal itself. To the earth-bound observer, the reasoning thus far may have been perfectly clear, hopefully at least, but—to speak for the drifting shark—it certainly must have sounded like gibberish. As stated above, to understand the challenges a shark faces, we must explore the situation as observed in the animal's frame of reference. But in the shark's frame, the animal and its ampullary sense organs are not moving. Since in its own frame the animal's velocity $\mathbf{v}'$ is zero, also its motionally induced electric field $\mathbf{v}' \times \mathbf{B}'$ is zero. Without moving, it cannot even sense the presence of the earth's magnetic field, unless it has in addition to its electric sense an independent, non-induction-based magnetic sense. If neither the stream nor the drifting animal is subject to a motionally induced electric field $\mathbf{v}' \times \mathbf{B}'$, how might the animal sense the strength and upstream direction of its drift in an ocean stream?

Rest assured, the view of the earth-bound observer is completely consistent with the inferred perception by the drifting shark, as classically follows from the Liénard-Weichert postulation of retarded potentials and, in modern physics, from Einstein's principle of special relativity. Both treatises come to the conclusion that moving in an ambient *magnetic* field in the earth frame of reference is equivalent to being exposed to an ambient *electric* field E'_{amb} in the shark's frame of reference. The imposed E'_{amb} ambient electric field not only nulls the $\mathbf{v}' \times \mathbf{B}'$ of the whole world outside the shark and the stream, which is moving with a velocity $\mathbf{v}' = -\mathbf{v}$, but it also penetrates the parts that are at rest, that is, the stream and the shark. In that sense, sharks do not directly experience a magnetic field per se but orient to the electric field E'_{amb} ensuing from the, relative to the shark and the stream, moving electric current loops deep inside the earth that, we assume, create the earth's magnetic field. In the shark frame, the current loops appear charged depending on the strength and direction of the relative velocity $\mathbf{v}'$ between the current loops and the stream plus the animals. Hence, the ambient electric field E'_{amb} in the shark frame corresponds to the $\mathbf{v} \times \mathbf{B}$ in the earth frame. The shark need not and cannot know where the E'_{amb} comes from, but it evidently has learned to appreciate it just as well as we—as scientists—appreciate the $\mathbf{v} \times \mathbf{B}$, both of which provide the same orientational information. Note: the E'_{amb} is not a purely electrostatic field, as it results from moving charges, hence need not be curl-free, just as much as $\mathbf{v} \times \mathbf{B}$ needs not be curl free, see Appendix 4.

The pleasing corollary of the relativistic view is the fact that, even though the driving forces and the induced electric field $\mathbf{v} \times \mathbf{B}$ in the earth frame and the imposed electric field E'_{amb} in the shark frame seem to be of different natures, yet the resulting $-\rho\mathbf{J}$ and $-\rho\mathbf{J}'$ fields are the same from both points of view except for the negligible relativistic difference at low speed between the current densities $\mathbf{J}$ and $\mathbf{J}'$ in the two frames. The same applies to the $\mathbf{B}$ and $\mathbf{B}'$. Therefore, the primes will be omitted, except for the $\mathbf{v}'$ and E'_{amb}. Hence, expressed in terms of $-\rho\mathbf{J}$, the physically adequate stimulus is essentially the same in either frame of reference, as Einstein pointedly noted in his original 1905 article. Studying the underwater world from the shark's point of view is highly recommended for making clear what the animal does and does not detect, hence on what sensory basis it makes its decisions. True, for this, the special theory of relativity is not strictly needed, since Faraday's law, which is a limiting consequence of the special theory of relativity, suffices at the low speeds here considered. Interestingly though, Einstein started the special theory of relativity with a fictive example equivalent to a world with its magnetic field at rest and the shark moving versus a world with its magnetic field moving and the shark at rest. Clearly, electric and magnetic fields are relative concepts. By detecting a motional-electric field in their own frame of reference, the animals orient to—what we in our frame of reference call—the earth's magnetic field.

I learned the pertinent physics in the first instance from the first edition of Edward M. Purcell's "Electricity and Magnetism" textbook. I am proud in later years to have shown and discussed my stingray motion-electric orientation experiments with him at the Woods Hole Oceanographic Institution. At the time, he was working on his 3rd edition, which after his death was completed and extended by David J. Morin. I strongly recommend the chapter on the motional electric principle of the 3rd edition, which gives an excellent relativistic introduction to a metal rod moving in a magnetic field. To make the step from a metal rod moving in vacuum to a seawater ocean stream moving in a seawater medium, I heavily relied on the scholarly article of Longuett, Stern, and Stommel, 1954, my senior Woods Hole colleagues of later years. Their classic mathematical-physics treatise served as a starting point for me to expand the theory of motional electric induction to a shark moving in a seawater medium in the presence of the earth's magnetic field.

2.2 Active Mode of Electro-orientation: Shark Detecting Compass Heading

2.2.1 *Physical Principles of Active Electro-orientation: Forces on Ionic Charge Carriers*[3]

In discussing a shark traveling in still open-ocean waters, thereby *actively* interacting with the earth's magnetic field, we will closely follow, almost verbatim, the same line of reasoning as in the instance of a shark drifting in an open-ocean stream *passively* interacting with the earth's magnetic field. This will render the discussions of a shark sensing its magnetic compass heading and a shark sensing its drift in ocean streams largely independent of each other, as well as clearly exhibit the close analogy between the two. Indeed, they are based on the very same physical principles, although they differ significantly in the specifics. The magnetically orienting shark relies predominantly on the horizontal component of the earth's magnetic field, whereas a shark drifting in an ocean stream relies mainly on the vertical magnetic component. Moreover, the charge accumulations on the skin of the shark are of different origin and consequence. In the magnetically orienting shark, the charges have an inhibiting effect, whereas in the drifting shark they have in general an enhancing effect. The fresher the surrounding medium, the more the inhibiting effect is in the magnetically orienting animal and the less the enhancing effect is in the drifting animal. Hence, the conditions are most favorable for marine sharks, less favorable for freshwater fish, and dismal for airborne birds.

2.2.1.1 Shark Traveling in Earth's Magnetic Field (Fig. 2.5)

Imagine a shark moving at a constant, horizontal, rectilinear velocity $\mathbf{v}$ relative to the water, assuming—for now—the water to be at rest in the earth frame of reference. What information might the animal derive from the motionally induced electric field $\mathbf{v} \times \mathbf{B}_h$ it actively generates by interacting with the horizontal component $\mathbf{B}_h$ of the earth's magnetic field? Let us again first ask a scientist at rest in the earth frame of reference, who observes the shark and its ampullary sense organs move in the earth's magnetic field, and then a shark in its own, proper frame of reference, in which the animal and, hence, its ampullary sense organs are at rest. With $\mathbf{v}$ and $\mathbf{B}_h$ horizontally directed, the vector product $\mathbf{v} \times \mathbf{B}_h$ points upwards in a shark moving in an easterly direction, as shown in the vector diagram of Fig. 2.5. We will assume the velocity of the stream $\mathbf{v}$ and the magnetic field $\mathbf{B}_h$ to be uniform and independent of time, the shark to be ellipsoidal in shape, and the skin and body tissues to have an average homogeneous resistivity ρ, about a hundred times higher than the

[3] The author recommends consulting Appendix 2 now, or to come back to it at any time later, to make sure the reader understands what the mathematical symbols physically stand for.

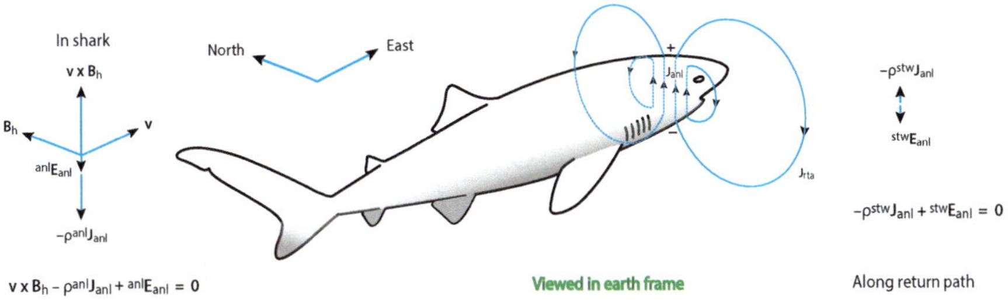

Fig. 2.5 Shark traveling ocean waters heading due east with velocity **v** creates, by interacting with horizontal component of the earth's magnetic field $\mathbf{B}_h$, vertically directed, *motionally induced electric field* $\mathbf{v} \times \mathbf{B}_h$. Resulting *induced electric current* of density ${}^{anl}\mathbf{J}_{anl}$ accumulates – and + surface charges on ventral and dorsal sides of animal, causing *motional electric field* ${}^{anl}\mathbf{E}_{anl}$. Left: ${}^{anl}\mathbf{J}_{anl}$, caused by $\mathbf{v} \times \mathbf{B}_h$ is impeded by *ohmic force* $-\rho^{anl}\mathbf{J}_{anl}$, with ρ resistivity of tissues, including skin, and by motional electric field ${}^{anl}\mathbf{E}_{anl}$ of – and + surface charges on ventral and dorsal sides of animal, hence, $\mathbf{v} \times \mathbf{B}_h - \rho^{anl}\mathbf{J}_{anl} + {}^{anl}\mathbf{E}_{anl} = 0$. Right: electric field of + and – surface charges furthermore drives electric current of density ${}^{stw}\mathbf{J}_{anl}$ along return paths through still seawater, along which $-\rho^{stw}\mathbf{J}_{anl} + {}^{stw}\mathbf{E}_{anl} = 0$. Note that $\mathbf{v} \times \mathbf{B}_h$, $-\rho^{anl}\mathbf{J}_{anl}$, $-\rho^{stw}\mathbf{J}_{anl}$, ${}^{anl}\mathbf{E}_{anl}$, and ${}^{stw}\mathbf{E}_{anl}$ denote the forces per unit positive charge. Lengths of arrows measure body-high, pore-to-pore averages of *forces per unit charge on positive ions*. To obtain the forces on the individual ions, they must be multiplied by the charges of the ions

surrounding seawater. The horizontal, motionally induced electric field $\mathbf{v} \times \mathbf{B}_v$ resulting from the shark's interaction with the vertical component of the earth's magnetic field $\mathbf{B}_v$ will be dealt with in a later paragraph. How the animals detect the all-important DC electric fields with their low-frequency AC receptors will be explained in Sect. 3.2. For now, we discuss the motional-electric signals the animal receives when moving at a constant rectilinear velocity, that is, gliding at a velocity **v** though the water, and the information that the signals offer the animal. The constraints will be relaxed later on.

2.2.1.2 In the Shark (Figs. 2.5 and 2.6)

The vertically directed motionally induced electric field $\mathbf{v} \times \mathbf{B}_h$ exerts an electric force on the ionic charge carriers in the shark, which outside the ampullary sense organs lead to a steady, induced electric current of density ${}^{anl}\mathbf{J}_{anl}$ flowing in the eastbound shark ventro-dorsally through the body, with the positive ions traveling in the direction of $\mathbf{v} \times \mathbf{B}_h$ and the negative ions in the opposite direction. At the skin, where the velocity **v**, hence $\mathbf{v} \times \mathbf{B}_h$ changes steeply, the ions entering the animal accelerate, whereas the ions leaving the animal decelerate, thus accumulating – and + ionic charges on the animal's ventral and dorsal surfaces. As a result, the electric current across the shark is impeded by two forces: the ohmic force $-\rho^{anl}\mathbf{J}_{anl}$, with ρ the resistivity of the skin, tissues, and body fluids, and the motional electric field ${}^{anl}\mathbf{E}_{anl}$ of the – and + charges on the ventral and dorsal skin surfaces.

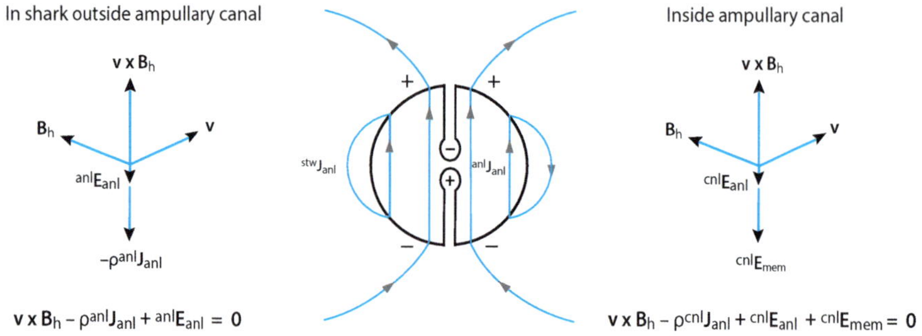

Fig. 2.6 Dynamic equilibrium of forces acting on positive ion charges in traveling shark. Middle: cross-section of shark with differential pair of ventro-dorsal ampullary sense organs and ampullae proper enclosed in ampullary capsule; – + surface charges on skin; + – surface charges on epithelia lining ampullary canals and blind ampullae proper; current of density ${}^{anl}\mathbf{J}_{anl}$ flowing across animal, though not through ampullary sense organs; current of density ${}^{stw}\mathbf{J}_{anl}$ returning through surrounding water. Left: vector diagram inside shark, but outside ampullary sense organs. Right: vector diagram inside ampullary canals. Lengths of arrows measure *forces per unit charge on positive ions*

The surface charges on the skin furthermore drive an electric current of density ${}^{stw}\mathbf{J}_{anl}$ along return paths through the still seawater. The surface-charge-increasing current of density ${}^{anl}\mathbf{J}_{anl}$ across the shark and the surface-charge-diminishing current of density ${}^{stw}\mathbf{J}_{anl}$ along the return paths through the seawater adjust the – and + surface charges on opposite skin surfaces such as to render the current continuous around the closed circuit through the animal and the surrounding seawater (Fig. 2.6). Note that the minus sign in $-\rho^{anl}\mathbf{J}_{anl}$ is part of the symbol, indicating that the direction of $-\rho^{anl}\mathbf{J}_{anl}$ is opposite to that of the current density ${}^{anl}\mathbf{J}_{anl}$. Note also that ${}^{anl}\mathbf{E}_{anl}$ in the animal and ${}^{stw}\mathbf{E}_{anl}$ in the still water apply to different regions of the same field, to wit, the field of the charge accumulations at opposite sides of the animal. The two regions differ in that the shark is in motion and is characterized by the presence of $\mathbf{v} \times \mathbf{B}_h$, whereas the return path remains still and is characterized by the absence of $\mathbf{v} \times \mathbf{B}_h$.

In the shark, but outside the ampullary sense organs, the motionally induced electric force $\mathbf{v} \times \mathbf{B}_h$, the ohmic force $-\rho^{anl}\mathbf{J}_{anl}$, and the electric force ${}^{anl}\mathbf{E}_{anl}$ of the charge accumulations are in dynamic equilibrium, $\mathbf{v} \times \mathbf{B}_h - \rho^{anl}\mathbf{J}_{anl} + {}^{anl}\mathbf{E}_{anl} = 0$, whereas along the return paths through the still water, where $\mathbf{v}$ and, hence, $\mathbf{v} \times \mathbf{B}_h$ are zero, solely the $-\rho^{stw}\mathbf{J}_{anl}$ and ${}^{stw}\mathbf{E}_{anl}$ are in dynamic equilibrium, $-\rho^{stw}\mathbf{J}_{anl} + {}^{stw}\mathbf{E}_{anl} = 0$, not taking into account the motion of the water displaced by the moving animal. Dynamic equilibrium means that the positive and negative ions have reached their limiting velocities after a brief transient period of acceleration, during which the velocity-dependent ohmic $-\rho^{anl}\mathbf{J}_{anl}$ and the motional electric field ${}^{anl}\mathbf{E}_{anl}$ of the – and + surface charges accumulating on the ventral and dorsal surfaces of the animal have grown so strong as to balance the motionally induced electric field $\mathbf{v} \times \mathbf{B}_h$. The amount of surface charges on the ventral and dorsal sides of the animal depends on how readily the charges are carried off along the return path through the surrounding seawater, hence, on the resistance of the return paths. After the stationary state has been reached, no further changes occur at constant **v**, $\mathbf{B}_h$, and ρ. Moreover, in our circular-cross-section, ellipsoidal, homogenous resistivity model of the shark, not only the $\mathbf{v} \times \mathbf{B}_h$, but also the $-\rho^{anl}\mathbf{J}_{anl}$ and ${}^{anl}\mathbf{E}_{anl}$ are inherently uniform.

2.2.1.3 In the Ampullary Sense Organs (Fig. 2.7)

In the ampullary sense organs and the capsules in which the ampullae proper are embedded, the motionally induced electric field $\mathbf{v} \times \mathbf{B}_h$ and the oppositely directed motional electric field $\mathbf{E}_{anl}$ of the charges accumulated on the shark's ventral and dorsal skin surfaces are parts of the same fields as those in the shark's body tissues. However, since the epithelial membranes lining the gel-filled ampullary canals and blind ampullae proper are purely dielectric, the current density $\mathbf{J}_{cnl}$ in the ampullary canals, driven by the $\mathbf{v} \times \mathbf{B}$ and the $\mathbf{E}_{anl}$, ceases to exist after a very short transient current accumulating + and – charges on the thin epithelial walls of the blind ampullary organs, in particular on the receptor-cell membranes of the sensory epithelia

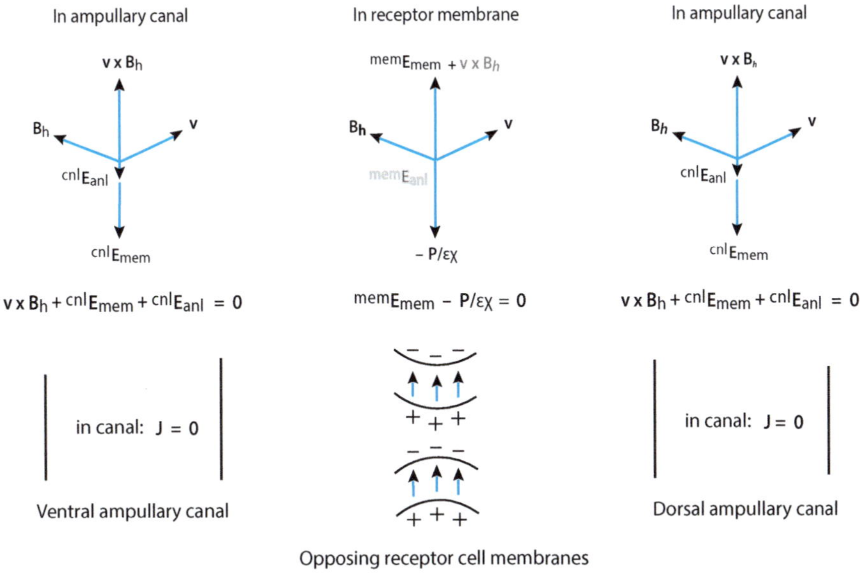

Fig. 2.7 Forces on positive ions in ampullary canals and receptor cell membranes. Left and right: since in ampullary organs total current density **J** and, hence, total ohmic field $-\rho\mathbf{J}$ are zero, motionally induced electric field $\mathbf{v} \times \mathbf{B}_h$, motional electric field ${}^{cnl}\mathbf{E}_{anl}$ of surface charges at ventral and dorsal skin surfaces of animal, and electric field ${}^{cnl}\mathbf{E}_{mem}$ of charges on thin epithelial lining of ampullary canal and ampulla proper are in dynamic equilibrium, $\mathbf{v} \times \mathbf{B}_h + {}^{cnl}\mathbf{E}_{anl} + {}^{cnl}\mathbf{E}_{mem} = 0$. Middle: In sensory membranes of ampullae proper, ${}^{mem}\mathbf{E}_{mem}$ field of charges on epithelial lining of ampullae proper, is orders of magnitude stronger than $\mathbf{v} \times \mathbf{B}_h + {}^{cnl}\mathbf{E}_{anl} + {}^{cnl}\mathbf{E}_{mem}$, due to thinness of receptor membrane compared to length of ampullary canal. Grayed-out quantities negligibly small. Hence, in receptor membrane, ${}^{mem}\mathbf{E}_{mem}$ force on ions, including gating charges of voltage-sensitive ion channels, is for all practical purposes balanced by $-{}^{mem}\mathbf{P}/\varepsilon\chi_{mem}$. Hence, ${}^{mem}\mathbf{E}_{mem} - {}^{mem}\mathbf{P}/\varepsilon\chi_{mem} = 0$

lining the ampullae proper. As a result, the electric field $\mathbf{E}_{cnl}$ of the accumulated epithelial charges nulls the $\mathbf{v} \times \mathbf{B}_v$ and the $\mathbf{E}_{anl}$, causing the $\mathbf{J}_{cnl}$ transient quickly to vanish. Hence, in the stationary state $\mathbf{v} \times \mathbf{B}_h + \mathbf{E}_{anl} + \mathbf{E}_{cnl} = 0$, for now not taking the excitation of the sensory epithelia in the wall of the blind ampullae proper into account. Since the $\mathbf{v} \times \mathbf{B}_v$ in our model is virtually uniform, the $\mathbf{E}_{anl}$ and $\mathbf{E}_{cnl}$ fields, when summed, are in the ampullae and the ampullary canals uniform as well, as a consequence of the self-adjusting charge distribution on the wall of the canals and ampullae proper (Fig. 2.7, left and right).

So far, the epithelial membranes lining the ampullary canals and the blind ampullae proper were assumed to be purely dielectric. In the live animal, though, the lobed ampullae proper contain the sensory maculae, which, being electrically excitable, are not purely dielectric, but feature a *negative resistance*. However, here we will first consider the membranes of the maculae to be perfectly dielectric, though they actively produce negative input currents exiting to the outside through the ampullary canals, in response to pore-positive stimuli, the consequences of which will be detailed in Sect. 2.3.2. Next considering the sensory membranes excitable will change the situation, but not invalidate the physical reasoning. This two-step approach of first considering the purely physical currents and then adding the physiological currents ties together the physical approach with the more recent findings concerning the amplifying physiological operation of the receptor cells.

2.2.1.4 In the Receptor-Cell Membranes (Fig. 2.6)

As the surface charges nulling the ampullary canal currents accumulate on the side of the receptor-cell membranes facing the ampullary gel, necessarily equal amounts of surface charges of opposite polarity accumulate on the other, the capsular-stroma side of the receptor-cell membranes. The membranes thus form ultra-thin, electrolytic capacitors, in which, by geometric design, the electric ${}^{mem}\mathbf{E}_{mem}$ field of the ampullary charges is orders of magnitude stronger than the electric ${}^{cnl}\mathbf{E}_{cnl}$ fields in the conductive, gelatinous contents of the ampullary canals and ampullae proper. Note that the $\mathbf{E}_{mem}$ field and $\mathbf{E}_{cnl}$ field, respectively inside and outside the receptor membrane, refer to two different regions of the same electric field produced by the same charges. The resulting dense concentrations of surface charges subject the ultra-thin receptor membranes to such strong electric fields that the $\mathbf{v} \times \mathbf{B}_v$, ${}^{mem}\mathbf{E}_{str}$, and ${}^{mem}\mathbf{E}_{anl}$ in the membranes are negligibly small compared to the ${}^{mem}\mathbf{E}_{mem}$. Hence, even though the charges on the receptor membranes result from the body-wide $\mathbf{v} \times \mathbf{B}_v$, ${}^{anl}\mathbf{E}_{str}$, and ${}^{anl}\mathbf{E}_{anl}$ fields, within the ultra-thin membranes, it is almost exclusively the orders-of-magnitude stronger ${}^{mem}\mathbf{E}_{mem}$ field that acts on the bound molecular charges and the gating charges of the voltage-sensitive ion channels.

In the receptor membranes, the bound molecular charges that the geometrically greatly amplified ${}^{mem}\mathbf{E}_{mem}$ field acts upon are held back by the dielectric polarization reactance force $-{}^{mem}\mathbf{P}/\varepsilon\chi$, with $\mathbf{P}$ the polarization density, ε the electric permittivity, and χ the electric susceptibility. Thus, since in the membranes the $\mathbf{v} \times \mathbf{B}_v$, ${}^{mem}\mathbf{E}_{str}$, and

${}^{mem}\mathbf{E}_{anl}$ are vanishingly small compared to the ${}^{mem}\mathbf{E}_{mem}$, the space-averaged ${}^{mem}\mathbf{E}_{mem}$ is for all intents and purposes in dynamic equilibrium with the membrane's molecular reactance force $-{}^{mem}\mathbf{P}/\varepsilon\chi$. Hence, practically, we may reckon ${}^{mem}\mathbf{E}_{mem}-{}^{mem}\mathbf{P}/\varepsilon\chi = 0$. By acting on the gating chargers of the ion channels, these ultimately are the forces responsible for the detection of the electrical signals (Figs. 2.6, right and 2.7, middle). The $-{}^{mem}\mathbf{P}/\varepsilon\chi$ restraining the bound ionic charges in the receptor-cell membranes represent the functional counterpart of the $-\rho^{anl}\mathbf{J}_{anl}$ restraining the ionic charges moving through the animal's body tissues outside the ampullary sense organs. How we arrive at this important conclusion will be explained in the next section.

2.2.2 *Physical Principles of Active Electro-orientation: In Terms of Voltages*[4]

How strong is the electrical excitation field $\mathbf{E}_{mem}$ in the receptor-cell membranes of a shark traveling in the earth's magnetic field? To gauge $\mathbf{E}_{mem}$, we will express the forces in terms of voltages and potential differences, that is, *as their line integrals along paths from one point to another*, rather than expressing the forces in terms of dynamic equilibria at individual points.

2.2.2.1 Governing Equations

In mathematical physics, electric and magnetic field equations can be expressed in differential form in terms of *forces* or in integral form in terms of *voltages*. The two modes of expression give both penetrating yet different insights, hence they are complementary. It must be well understood that $\mathbf{v} \times \mathbf{B}$ and $-\rho\mathbf{J}$ are gradients of voltages, as they tend to increase the voltage, whereas $\mathbf{E}$ is the *negative* of the gradient of the voltage, as it tends to decrease the voltage. The perilous habit of not consistently adhering to either the force approach or the voltage approach may lead to great confusion. I have struggled enough with the problem to feel urged to present the two approaches side by side, yet each on its own to avoid misapprehensions.

In terms of the *forces* governing the motional electric field a shark drifting in an open-ocean stream receives, the stationary situation is ruled by the equation $\mathbf{v} \times \mathbf{B} - \rho\mathbf{J} + \mathbf{E} = 0$, that is, the driving force $\mathbf{v} \times \mathbf{B}$, the resistive force $-\rho\mathbf{J}$, and the

[4] At this stage, the reader is advised to review the mathematical concepts of *divergence* and *curl*, the two properties that determine the physical nature of the fields and help explain the difference between the concepts of *voltage gradients* and *potential gradients*. See Appendices 3 and 4. Though not required, the present section on active orientation will greatly benefit from first reading the preceding section on passive orientation, following the author's original line of thought, from electrodes and sharks drifting in ocean streams to electrodes and sharks actively moving through the water.

electrostatic force **E** of the – and + charges accumulated by the electric current of density **J** are in dynamic equilibrium. In the stream, the $\mathbf{v} \times \mathbf{B}$ and **J** are in the positive → or ↑ direction in our examples; whereas the $-\rho\mathbf{J}$ and **E** are in the negative ← or ↓ direction. When **E** approaches nil, $\mathbf{v} \times \mathbf{B}$ and $-\rho\mathbf{J}$ tend to a steady equilibrium, that is, in the limit they are of the same strength but opposite in direction, hence, $\mathbf{v} \times \mathbf{B} - \rho\mathbf{J} = 0$. However, when **E** is not nil due to + and – charge accumulations at the lateral boundaries of the stream, $\mathbf{v} \times \mathbf{B} - \rho\mathbf{J} + \mathbf{E} = 0$. Hence, **E** causes the strength of $-\rho\mathbf{J}$ to diminish in not perfectly loaded ocean streams. These intricate force-based relationships were the topic of the previous section.

In terms of *voltages*, given that the electric force **E** equals the negative of the voltage gradient $\nabla\varphi$, the steady situation is governed by the *integral* equation $\int \mathbf{v} \times \mathbf{B} \cdot \mathbf{ds} + \int -\rho\mathbf{J} \cdot \mathbf{ds} = \int \nabla\varphi \cdot \mathbf{ds}$, that is, the electrical energy imparted $\int \mathbf{v} \times \mathbf{B} \cdot \mathbf{ds}$ plus the negative of the electrical energy dissipated $\int -\rho\mathbf{J} \cdot \mathbf{ds}$ equals the electrical energy $\int \nabla\varphi \cdot \mathbf{ds}$ stored into the electric field of the + and – charges accumulated by the current of density **J**. When $\int \nabla\varphi \cdot \mathbf{ds}$ approaches nil, the imparted energy tends to equal the energy dissipated, that is, in the limit $\int \mathbf{v} \times \mathbf{B} \cdot \mathbf{ds} + \int -\rho\mathbf{J} \cdot \mathbf{ds} = 0$, rendering the physically adequate stimulus $\int -\rho\mathbf{J} \cdot \mathbf{ds}$ equal to $-\int \mathbf{v} \times \mathbf{B} \cdot \mathbf{ds}$. However, when $\int \nabla\varphi \cdot \mathbf{ds}$ is positive, $\int \mathbf{v} \times \mathbf{B} \cdot \mathbf{ds} + \int -\rho\mathbf{J} \cdot \mathbf{ds} = \int \nabla\varphi \cdot \mathbf{ds}$, part of the electrical energy is stored, increasing $\int \nabla\varphi \cdot \mathbf{ds}$ by the separation of – and + charges, thereby reducing the energy dissipated by $\int -\rho\mathbf{J} \cdot \mathbf{ds}$ and rendering the physically adequate stimulus $\int -\rho\mathbf{J} \cdot \mathbf{ds}$ less than $-\int \mathbf{v} \times \mathbf{B} \cdot \mathbf{ds}$, as in not perfectly loaded ocean streams. These energy-based relationships are the topic of the present section.

All very well, but I must reiterate: the limit of zero charge accumulations, hence zero **E**, can never be reached, since the seawater return path does not have zero resistance. Common lack of awareness concerning this fundamental issue has, in the scientific literature, led to an erroneous oversimplified version of my original conception, one of the main conundrums to be clarified in Sect. 2.3. As Einstein advocated: "As simple as possible, but not simpler."

Since in the traveling shark and the still water, **v** and $\mathbf{B}_h$ are—at least for now—assumed to be constant and uniform, hence free of curl, $\mathbf{v} \times \mathbf{B}_h$, $-\rho^{anl}\mathbf{J}_{anl}$ and $-\rho^{stw}\mathbf{J}_{anl}$ are *potential gradients*, pointing energetically uphill. However, the ${}^{anl}\mathbf{E}_{anl}$ and ${}^{stw}\mathbf{E}_{anl}$, inherently are the *negatives* of the potential gradients ${}^{anl}\nabla\varphi_{anl}$ and ${}^{stw}\nabla\varphi_{anl}$, where ${}^{anl}\mathbf{E}_{anl} = -{}^{anl}\nabla\varphi_{anl}$ is the field of the – and + surface charges on the skin of the animal measured inside the animal, ${}^{stw}\mathbf{E}_{anl} = {}^{stw}\nabla\varphi_{anl}$ is the field of the same surface charges on the skin of the animal measured along the return path through the still water. Note that the charge accumulations on the skin of the animal have very different origins in the shark swimming in still water and in the drifting shark (be fully aware of this!).

Similarly, in ampullary sense organs, the $-\mathbf{P}/\varepsilon\chi$ is a voltage gradient. Furthermore, ${}^{cnl}\mathbf{E}_{str} = -{}^{cnl}\nabla\varphi_{str}$ is the field of the space charges at the lateral boundaries of the stream measured in the ampullary canal, ${}^{cnl}\mathbf{E}_{anl} = -{}^{cnl}\nabla\varphi_{anl}$ the field of the charges on opposite skin surfaces in the ampullary canal measured in the canal, ${}^{cnl}\mathbf{E}_{mem} = -{}^{cnl}\nabla\varphi_{mem}$ the field of the charges on opposite sides of the receptor-cell membranes measured in the ampullary canal, ${}^{mem}\mathbf{E}_{mem} = -{}^{mem}\nabla\varphi_{mem}$ the field resulting from the charges on opposite sides of the receptor-cell membranes measured inside the receptor-cell membranes. The ρ denotes the resistivity at the point of observation or along the path of integration, either in the ocean or in the animal. The importance of focusing on the fields that are free of curl, hence conservative potential gradients, is the characteristic that their pore-to-pore line integrals are path-independent, keeping mathematics initially uncluttered for the sake of clarity.

At the skin surfaces of the shark, where **v** changes steeply in the *parallel* direction, hence charges accumulate, $\mathbf{v} \times \mathbf{B}_h$ is not free of divergence. At the skin surfaces, where **v** changes steeply in the direction *normal* to **v**, as at the bottom of the stream, $\mathbf{v} \times \mathbf{B}_v$ is not free of curl, hence a voltage gradient, not a potential gradient. These issues will be discussed later in greater detail, when we will relax the constraints. Until then, we will not cross, but stay within the boundaries of the animal, either entirely within the animal or entirely within the waters outside the animal, where the divergence and the curl are zero, assuming the velocity **v** of the animal, the magnetic field **B**, and the resistivity ρ of the water and the animal to be uniform and time-independent.

2.2.2.2 Introduction and Preview of Analysis

To illustrate the governing equations in their simplest form, we will in the instance of electrodes towed through still water analyze the motional electric measurements made with long salt bridges connected via non-polarizable electrodes to an ideal voltmeter, by definition having infinite input resistance, hence not drawing any electric current. The salt bridges will be labeled *cnl* as they are the equivalents of the sharks' ampullary canals; the electrodes & sensing elements will be labeled *mem* as they are the equivalents of the sharks' electroreceptor membranes. The salt bridges and electrodes are assumed to have such slender shape and small volume that they neither significantly alter the motion of the water nor the imposed electric current fields due to electrical charges accumulating on opposite surfaces. The differential pairs of salt-bridge pores will be oriented vertically for interaction with the horizontal component of the earth's magnetic field and horizontally for interaction with the vertical component of the earth's magnetic field.

We will analyze the same situation for a shark with a pair of voltage-sensitive ampullae of Lorenzini, this time taking also the animal's body, in which the ampullae are embedded, into account. The shark's body is large and of quite different resistivity than the surrounding seawater, so that the imposed electric current fields cause significant charge accumulations on opposite sides of the animal, adding a distributed dipole current field. The amount of charge accumulations depends on the average resistivity of the shark and the resistivity of the surrounding seawater. As for the effect of the shark's body on the motion of the water, we will estimate the necessary corrections below.

First, we will preview the general rules governing (1) the electric fields physical oceanographers measure at sea with towed salt-bridge electrodes connected to a voltmeter, and (2) traveling sharks measure with their electroreceptors, the ampullae of Lorenzini. We will list the idealized situations considered thus far.

Sharks may similarly reset their electroreceptors by not accelerating or decelerating for a time long compared to their time constant of adaptation.

~ Electrodes Towed Through Still Water

The volume of water is *still* and devoid of electric fields. The *moving* electrodes & voltmeter measure the line integral of the *motionally induced* –**v** × **B** in the salt-bridge electrodes, leaving the seawater devoid of electric fields.

$$!!!\text{Ergo}: \int^{\text{mem}} \nabla\varphi_{\text{mem}} \cdot \mathbf{ds} = -\int \mathbf{v}\times\mathbf{B}\cdot\mathbf{ds}$$

~ Electrodes Drifting and Towed in Stream

$$!!!\text{Ergo}: \int^{\text{mem}} \nabla\varphi_{\text{mem}} \cdot \mathbf{ds} = \int -\rho^{\text{str}}\mathbf{J}_{\text{str}} \cdot \mathbf{ds} - \int \mathbf{v}_{\text{t}} \times \mathbf{B}_{\text{v}} \cdot \mathbf{ds}$$

~ Shark Traveling Through Still Water The volume of water is *still* and devoid of electric fields. The ampullae of Lorenzini of the *moving* shark measure the line integral of the *resistive* –ρ**J** across the animal through the skin and body tissues bypassing the ampullary sense organs, which equals the line integral of the motionally induced –**v** × **B** moving electrodes detect, *plus* the line integral of the motional electric field ∇*φ* of the charges accumulation on opposite skin surfaces.

$$!!!\ \text{Ergo}: \int^{\text{mem}} \nabla\varphi_{\text{mem}} \cdot \mathbf{ds} = \int^{\text{mem}} \nabla\varphi_{\text{mem}} \cdot \mathbf{ds} = \int -\rho^{\text{anl}}\mathbf{J}_{\text{anl}} \cdot \mathbf{ds}$$

~ Shark Drifting and Traveling in Stream

$$!!!\ \text{Ergo}: \int\left(-\rho^{\text{str}}\mathbf{J}_{\text{str}} - \rho^{\text{str}}\mathbf{J}_{\text{anl}}\right)\cdot\mathbf{ds} + \int -\rho^{\text{anl}}\mathbf{J}_{\text{anl}} \cdot \mathbf{ds}$$

In the following mathematical derivations, it is repeatedly stated that:

"Practically $-{}^{\text{mem}}\mathbf{P}_{\text{mem}}/\varepsilon\chi \cdot \mathbf{ds} = \int^{\text{mem}}\nabla\varphi_{\text{mem}} \cdot \mathbf{ds}$ and $\int({}^{\text{cnl}}\nabla\varphi_{\text{mem}} + {}^{\text{mem}}\nabla\varphi_{\text{mem}}) \cdot \mathbf{ds} = 0$."

$\int^{\text{mem}}\nabla\varphi_{\text{mem}} \cdot \mathbf{ds} = -{}^{\text{mem}}\mathbf{P}_{\text{mem}}/\varepsilon\chi \cdot \mathbf{ds}$ is virtually true since the $\int^{\text{mem}}\nabla\varphi_{\text{mem}}$ inside the sensory membrane is many orders of magnitude stronger than any other voltage gradients acting from the outside on the sensory membrane. $\int({}^{\text{cnl}}\nabla\varphi_{\text{mem}} + {}^{\text{mem}}\nabla\varphi_{\text{mem}}) \cdot \mathbf{ds} = 0$ is the very principle of voltmeters with their test leads, including the ampullae of Lorenzini with their gel-filled canals. Let us review why:

It must be well understood that the electric charges nulling the electric current in the salt bridges and in the ampullary canals are the same charges as those producing the electric fields in the voltage-sensitive receptor-cell membranes. The charges on the wall of the ampullary canal shaping the electric field in the canal do not enter the pore-to-pore line integral and need not be discussed any further. Hence, the salt-bridge and ampullary canal pore-to-pore integrals of ${}^{\mathrm{cnl}}\nabla\varphi_{\mathrm{mem}}+{}^{\mathrm{mem}}\nabla\varphi_{\mathrm{mem}}$ are nil. That is, the electrical charges on either side of the receptor membranes build up until the electric fields in terms of voltage gradients in the salt bridges or ampullary canals are nulled. Hence, all the voltages between the pores appear across the receptor membranes. That is the virtue of a true voltmeter, not to draw any current nor to add any voltage to the circuit. See also Appendix 5.

The shark travels in the positive direction from west to east → ; the magnetic field points positive up ↑ assuming we are in the southern hemisphere. The equations in the following mathematical discourse describe the various situations as they are observed in the earth frame of reference. How the equations transform to the shark frame of referenced will be explained below.

The mathematical details of the paragraphs marked with an "approximate" sign (~) may be glanced over, only to read the concluding lines marked with three exclamation marks (! ! !) and with a pound sign (#). Warning, to envision the various situations is challenging, especially when we introduce the shark into the stream. So that each situation can stand on its own, the in-between steps are spelled out every time again. The readers who will understand the following situations will find out that it all is rather simple in hindsight. Simple enough for sharks to practice the intricate physics.

Superscripts are omitted where it is obvious from the context where the pertinent physical quantity is measured.

The situations concerning electric fields measured technically with electrodes and by sharks with the ampullary system are self-contained and can be studied in isolation or in arbitrary order.

2.2.2.3 Detailed Mathematical Analysis

~ Electrodes Towed Through Still Water, Interaction with $\mathbf{B}_{\mathrm{h}}$, (a) Electric Field Vertical

In still water:

$$\mathbf{v}\times\mathbf{B}_{\mathrm{h}}=\mathbf{0}\ \text{and}\ -\rho^{\mathrm{stw}}\mathbf{J}_{\mathrm{stw}}=0\ \text{and}\ {}^{\mathrm{stw}}\mathrm{E}_{\mathrm{stw}}=0\qquad \text{forces on unit}^{+}\text{ions}$$

$$\mathbf{v}\times\mathbf{B}_{\mathrm{h}}=0\ \text{and}\ -\rho^{\mathrm{stw}}\mathbf{J}_{\mathrm{stw}}=0\ \text{and}\ {}^{\mathrm{stw}}\nabla\varphi_{\mathrm{stw}}=0\qquad \text{potential gradients}$$

Salt bridge pore-to-pore line integral along paths through still seawater:

$$\text{Path1}: \ V_{\text{p2}} - V_{\text{p1}} = \int\left(\mathbf{v}\times\mathbf{B}_{\text{h}} - \rho^{\text{stw}}\mathbf{J}_{\text{stw}}\right)\cdot\mathbf{ds} = \int^{\text{stw}}\nabla\varphi_{\text{stw}}\cdot\mathbf{ds} = 0 \quad \textit{no electric current!}$$

In towed salt bridges and in sensing elements, respectively: *no electric current*!

$$\mathbf{v}\times\mathbf{B}_{\text{v}} + {}^{\text{cnl}}E_{\text{mem}} = 0 \text{ and } -{}^{\text{mem}}\mathbf{P}_{\text{mem}}/\varepsilon\chi + {}^{\text{mem}}E_{\text{mem}} = 0 \qquad \text{forces on unit}^{+}\text{ions}$$

$$\mathbf{v}\times\mathbf{B}_{\text{v}} = {}^{\text{cnl}}\nabla\varphi_{\text{mem}} \text{ and } -{}^{\text{mem}}\mathbf{P}_{\text{mem}}/\varepsilon\chi = {}^{\text{mem}}\nabla\varphi_{\text{mem}} \qquad \text{potential gradients}$$

Salt bridge pore-to-pore line integral along paths through salt bridges and sensing elements:

$$\text{Path2}: \ V_{\text{p2}} - V_{\text{p1}} = \int\left(\mathbf{v}\times\mathbf{B}_{\text{v}} - {}^{\text{mem}}\mathbf{P}_{\text{mem}}/\varepsilon\chi\right)\cdot\mathbf{ds} = \int\left({}^{\text{cnl}}\nabla\varphi_{\text{mem}} + {}^{\text{mem}}\nabla\varphi_{\text{mem}}\right)\cdot\mathbf{ds}$$

Or, since practically $\int -{}^{\text{mem}}\mathbf{P}_{\text{mem}}/\varepsilon\chi \cdot \mathbf{ds} = \int {}^{\text{mem}}\nabla\varphi_{\text{mem}} \cdot \mathbf{ds}$ and $\int({}^{\text{cnl}}\nabla\varphi_{\text{mem}} + {}^{\text{mem}}\nabla\varphi_{\text{mem}}) \cdot \mathbf{ds} = 0$

$$\text{Path2}: \ V_{\text{p2}} - V_{\text{p1}} = \int\left(\mathbf{v}\times\mathbf{B}_{\text{h}} + {}^{\text{mem}}\nabla\varphi_{\text{mem}}\right)\cdot\mathbf{ds}$$

Comparing path 2 equation with path 1 equation:

$$\int\left(\mathbf{v}\times\mathbf{B}_{\text{h}} + {}^{\text{mem}}\nabla\varphi_{\text{mem}}\right)\cdot\mathbf{ds} = 0$$

$$\text{!!! Ergo vertically}: \int^{\text{mem}}\nabla\varphi_{\text{mem}}\cdot\mathbf{ds} = -\int\mathbf{v}\times\mathbf{B}_{\text{h}}\cdot\mathbf{ds} \qquad \downarrow \text{in negative direction}$$

In brief: since the line integral along pore-to-pore paths through the seawater is 0, the line integral along the pore-to-pore paths of integration through salt bridges & sensing elements is 0 as well, leaving the line integrals of ${}^{\text{mem}}\nabla\varphi_{\text{mem}} = -\mathbf{v}\times\mathbf{B}_{\text{h}} \downarrow$ for the voltmeter to detect.

Vertically oriented towed electrodes normal to the direction of towing thus measure, by interaction with the horizontal component of the earth's magnetic field, in still ocean water a motionally induced electric field $-\mathbf{v}\times\mathbf{B}_{\text{h}}$, which depends on the magnetic heading. The strength of $-\mathbf{v}\times\mathbf{B}_{\text{h}}$ is maximally negative ↓ when the electrodes are towed due east, maximally positive ↑ when the electrodes are towed due west, and zero when the electrodes are towed due north or due south.

~ Electrodes Towed Through Still Water, Interaction with $\mathbf{B}_{\text{v}}$, (b) Electric Field Horizontal

By analogy of the interaction with $\mathbf{B}_{\text{h}}$:

$$\text{!!! Ergo horizontally}: \int^{\text{mem}}\nabla\varphi_{\text{mem}}\cdot\mathbf{ds} = -\int\mathbf{v}\times\mathbf{B}_{\text{v}}\cdot\mathbf{ds} \qquad \leftarrow \text{in negative direction}$$

In brief: since the line integrals along pore-to-pore path through the seawater is 0, the line integrals along the pore-to-pore paths through salt bridges & sensing elements is 0 as well, leaving the line integrals of ${}^{\text{mem}}\nabla\varphi_{\text{mem}} = -\mathbf{v}\times\mathbf{B}_{\text{v}} \leftarrow$ for the voltmeter to detect.

Horizontally towed electrodes normal to the direction of towing thus measure, by interaction with the vertical component of the earth's magnetic field, in still ocean waters a motionally induced electric field $-\mathbf{v}\times\mathbf{B}_{\text{v}} \leftarrow$, which is independent of the magnetic heading.

Consistency

If in a stream drifting electrodes are pulled upstream so as to offset their drift, rendering them stationary, they measure the signal due to their drift *plus* the signal due to being towed upstream:

$$\int^{\text{mem}} \nabla\varphi_{\text{mem}}\cdot \mathbf{ds} = \left\{\int -\rho^{\text{str}}\mathbf{J}_{\text{str}}\cdot \mathbf{ds}\right\}\text{downstream drift} + \left\{\int \mathbf{v}_{\text{t}}\times\mathbf{B}\cdot\mathbf{ds}\right\}\text{upstream towed}$$
$$= \int^{\text{str}} \nabla\varphi_{\text{str}}\cdot\mathbf{ds}$$

as stationary electrodes in the stream indeed proved to measure.

~ Shark Traveling Through Still Water, Interaction with $\mathbf{B}_{\text{h}}$, (a) Electric Field Vertical

Inside traveling shark, but outside ampullary sense organs:

$$\mathbf{v}\times\mathbf{B}_{\text{h}} - \rho^{\text{anl}}\mathbf{J}_{\text{anl}} + {}^{\text{anl}}E_{\text{anl}} = 0 \qquad \text{forces on unit}^{+}\text{ions}$$

$$\mathbf{v}\times\mathbf{B}_{\text{h}} - \rho^{\text{anl}}\mathbf{J}_{\text{anl}} = {}^{\text{anl}}\nabla\varphi_{\text{anl}} \qquad \text{potential gradients}$$

The ${}^{\text{anl}}\mathbf{E}_{\text{anl}} = -{}^{\text{anl}}\nabla\varphi_{\text{anl}} \downarrow$ pertains to the electrostatic field due to the – and + charges on opposite skin surfaces of the animal. The ${}^{\text{anl}}\mathbf{E}_{\text{anl}}$ and $-\rho^{\text{anl}}\mathbf{J}_{\text{anl}}$ counter the $\mathbf{v}\times\mathbf{B}_{\text{h}}$.

Pore-to-pore line integrals along paths through skin & body tissues:

$$\text{Path1}: V_{\text{p2}} - V_{\text{p1}} = \int\left(\mathbf{v}\times\mathbf{B}_{\text{h}} - \rho^{\text{anl}}\mathbf{J}_{\text{anl}}\right)\cdot\mathbf{ds} = \int^{\text{anl}}\nabla\varphi_{\text{anl}}\cdot\mathbf{ds}$$

In ampullary canals and in receptor-cell membrane, respectively: *no electric current*!

$$\mathbf{v}\times\mathbf{B}_{\text{v}} + {}^{\text{cnl}}E_{\text{anl}} + {}^{\text{cnl}}E_{\text{mem}} = 0 \text{ and } -{}^{\text{mem}}\mathbf{P}_{\text{mem}}/\varepsilon\chi + {}^{\text{mem}}E_{\text{mem}} = 0 \quad \text{forces on unit}^{+}\text{ions}$$

$$\mathbf{v}\times\mathbf{B}_{\text{v}} = {}^{\text{cnl}}\nabla\varphi_{\text{anl}} + {}^{\text{cnl}}\nabla\varphi_{\text{mem}} \text{ and } -{}^{\text{mem}}\mathbf{P}_{\text{mem}}/\varepsilon\chi = {}^{\text{mem}}\nabla\varphi_{\text{mem}} \quad \text{potential gradients}$$

Pore-to-pore line integral along paths through sense organs:

$$\text{Path2}:\quad V_{p2}-V_{p1}=\int\left(\mathbf{v}\times\mathbf{B}_{v}-^{mem}\mathbf{P}_{mem}/\varepsilon\chi\right)\cdot\mathbf{ds}$$
$$=\int\left(^{cnl}\nabla\varphi_{anl}+^{cnl}\nabla\varphi_{mem}+^{mem}\nabla\varphi_{mem}\right)\cdot\mathbf{ds}$$

Or, since practically $\int -^{mem}\mathbf{P}_{mem}/\varepsilon\chi\cdot\mathbf{ds}=\int^{mem}\nabla\varphi_{mem}\cdot\mathbf{ds}$ and $\int(^{cnl}\nabla\varphi_{mem}+^{mem}\nabla\varphi_{mem})\cdot\mathbf{ds}=0$

$$\text{Path2}:\quad V_{p2}-V_{p1}=\int\left(\mathbf{v}\times\mathbf{B}_{h}+^{mem}\nabla\varphi_{mem}\right)\cdot\mathbf{ds}=^{cnl}\nabla\varphi_{anl}\cdot\mathbf{ds}$$

Comparing path 2 equation with path 1 equation:

$$\int\left(\mathbf{v}\times\mathbf{B}_{h}+^{mem}\nabla\varphi_{mem}\right)\cdot\mathbf{ds}=\int\left(\mathbf{v}\times\mathbf{B}_{h}-\rho^{anl}\mathbf{J}_{anl}\right)\cdot\mathbf{ds}$$

$$\text{!!! Ergo vertically}:\int^{mem}\nabla\varphi_{mem}\cdot\mathbf{ds}=\int-\rho^{anl}\mathbf{J}_{anl}\cdot\mathbf{ds}\qquad\downarrow\text{in negative direction}$$

In brief: the pore-to-pore path of integration through ampullary sense organs and the pore-to-pore path through animal but outside ampullary sense organs have the line integral of $\mathbf{v}\times\mathbf{B}_h$ and also the line integral of $^{anl}\nabla\varphi_{anl}$ due to charges on the skin in common, leaving line the integral of $-\rho^{anl}\mathbf{J}_{anl}$ for the ampullary sense organs of the traveling shark to detect.

A shark traveling in still ocean waters thus receives from interaction with the horizontal component of the earth's magnetic field a dorso-ventral ohmic $-\rho^{anl}\mathbf{J}_{anl}$ field ↓ resulting from the animal's motionally induced electric current flowing across the body. The field depends on the animal's magnetic heading. It is maximal plus or minus when the shark travels due east or west and minimal when the shark travels due north or south.

How the $-\rho^{anl}\mathbf{J}_{anl}$ field received by the animal relates to the driving motional electric field $\mathbf{v}\times\mathbf{B}_h$ depends on the degree of electrical loading of the shark by the seawater according to the relationship $\int^{anl}\nabla\varphi_{anl}\cdot\mathbf{ds}=\int(\mathbf{v}\times\mathbf{B}_h-\rho^{anl}\mathbf{J}_{anl})\cdot\mathbf{ds}$. Hence, $^{anl}\nabla\varphi_{anl}$, the motional electric field of the electrical charges on opposite skin surfaces specifies the vector difference between $-\rho^{anl}\mathbf{J}_{anl}$ and $\mathbf{v}\times\mathbf{B}_h$.

The $-\rho^{anl}\mathbf{J}_{anl}$ field is the stronger, the more heavily the shark is electrically loaded, that is, the smaller the charge accumulations on the opposite skin surfaces of the animal are, that is, the smaller the $^{anl}\nabla\varphi_{anl}$ is and the more $-\rho^{anl}\mathbf{J}_{anl}$ counteracts $\mathbf{v}\times\mathbf{B}_h$. Conversely, if the shark enters fresh waters, the charge accumulations become much larger, hence the $-\rho^{anl}\mathbf{J}_{anl}$ field becomes much weaker, and the animal measures only little. Hence, motion-electrically sensing magnetic compass headings is the more feasible the higher the resistance of the animal is relative to the surrounding medium. In that respect, ocean-dwelling sharks are among the most efficient.

Note: Although ideal towed electrodes do not create an electric field in the surrounding seawater, the moving shark, with its electroreceptors embedded in a large body, does. That is, the seawater $^{stw}\mathbf{E}_{anl}$ and $^{stw}\nabla\varphi_{anl}$ are not zero, unless the shark would electrically be completely shorted by the external medium, which it never is.

~ Shark Traveling in Still Water, Interaction with $\mathbf{B}_v$, (b) Electric Field Horizontal

$$\text{!!! Ergo horizontally: } \int^{mem} \nabla \varphi_{mem} \cdot \mathbf{ds} = \int -\rho^{anl} \mathbf{J}_{anl} \cdot \mathbf{ds} \qquad \leftarrow \text{in negative direction}$$

In brief: the pore-to-pore path of integration through the ampullary sense organs and the pore-to-pore path through animal but outside ampullary sense organs have $\mathbf{v} \times \mathbf{B}_h$ and $-\rho^{anl}\mathbf{J}_{anl}$ due to charges on the skin in common, leaving line integral of $-\rho^{anl}\mathbf{J}_{anl}$ for the ampullary sense organs of the traveling shark to detect.

A shark traveling in still ocean waters thus receives, by interaction with the vertical component of the earth's magnetic field, a horizontal ohmic $-\rho^{anl}\mathbf{J}_{anl}$ field ← resulting from the animal's motionally induced electric current flowing across the body. The field is independent of the animal's magnetic compass heading.

!!! Physically Adequate Stimulus

We thus arrive at the conclusion that, for a shark swimming in the open ocean, the ventro-dorsal $\int^{mem}\nabla\varphi_{mem} \cdot \mathbf{ds}$ equals $\int -\rho^{anl}\mathbf{J}_{anl} \cdot \mathbf{ds}$. Hence, *the pore-to-pore line integral of* $-\rho^{anl}\mathbf{J}_{anl}$, *the ohmic potential gradient resulting from the motionally induced electric current of density* ${}^{anl}\mathbf{J}_{anl}$ *through the animal's body tissues, measures the physically adequate stimulus the shark receives*. Note that the pore-to-pore line integral of $-\rho^{anl}\mathbf{J}_{anl}$ differs from the pore-to-pore line integral of $-\mathbf{v} \times \mathbf{B}_h$ by the line integral of ${}^{anl}\nabla\varphi_{anl}$ due to the charges on the ventral and dorsal skin surfaces of the animal.

To repeat for emphasis, we learn that, although the physically adequate stimulus results from the motionally induced electric field $\mathbf{v} \times \mathbf{B}_h$, it is not the motionally induced electric field $\mathbf{v} \times \mathbf{B}_h$ per se, nor the motional electric field ${}^{anl}\nabla\varphi_{anl}$ of the charge accumulations on the skin of the animal, but it is the resistive ohmic $-\rho^{anl}\mathbf{J}_{anl}$ field due to the electric current resulting from the motionally induced electric field. As the $-\rho^{anl}\mathbf{J}_{anl}$ grows larger, the charge accumulations on the skin due to the incomplete loading of the animal become less. For sharks in ocean waters the motional electric principle is nearly optimal, but for freshwater and airborne animals, it would be of less to no avail.

Consistency

If a shark drifting in an ocean stream swims upstream so as to offset its drift and to become stationary, it receives the signal due to its drift *plus* the signal due to swimming upstream: vertically, by interaction with the horizontal component of the earth's magnetic field.

The field of the charges accumulating on opposite skin surfaces of the sharks to divert the electric current of the stream around the animal, due to the shark being more resistive than the seawater, remains the same. The up-swimming shark in the down-flowing stream loads the stream.

$$\text{Down-stream}: \left(-\rho^{\text{anl}}\mathbf{J}_{\text{str}} - \rho^{\text{anl}}\mathbf{J}_{\text{anl}}\right)\cdot \mathbf{ds}$$

$$\text{Up-stream}: \int -\rho^{\text{anl}}\mathbf{J}_{\text{anl}}\cdot \mathbf{ds}\ \text{ plus a term due to loading of stream equals}: \int -\rho^{\text{anl}}\mathbf{J}_{\text{str}}\cdot \mathbf{ds}$$

This seemingly simple result has required from the author a lot of contemplation. He requires the math to be a succinct expression of his conceptual thinking. See for yourself whether he got it right.

By Extension

Sharks may passively drift in ocean streams flowing in any compass direction and in addition actively travel with respect to the water in any compass direction, hence receive many versions of the fields discussed so far in highly variable combinations. For us, that is not a problem, because all the fields discussed so far may simply be vectorially added according to the principle of superposition. How sharks may unravel the different situations by executing the proper exploratory movements will be the topic of the next section. Trust me, if we can do so, sharks must be able to do so much better.

Orders of Magnitude

To gauge—in a shark traveling ocean waters—the strength of the $\mathbf{E}_{\text{mem}}$ force acting on the gating charges of the voltage-sensitive ion channels of the receptor membranes, expressed in terms of the physically adequate stimulus $-\rho\mathbf{J}_{\text{anl}}$, let the distance between the skin pores of a differential pair of ampullae be 20 cm, and let the two opposing sensory epithelia—for now—consist of just two single membranes with a combined thickness of 10 nm. Hence, at an average $-\rho\mathbf{J}$ threshold field of 5 nV/cm in the body tissues outside the ampullary sense organs, a value proven to be readily detectable by the animal, the voltage difference between the 20 cm apart skin pores measures 100 or 50 nV for each of the two sensory membranes. However, since the sensory membranes are ≈200,000,000 × thinner than the ampullary canals are long, the voltage gradient in the sensory membranes is 5 nV/cm × 200,000,000 = 1 V/cm. For a traveling shark heavily loaded by the surrounding seawater, the averaged 5 nV/cm in the body tissues corresponds to a $-\rho\mathbf{J}_{\text{anl}}$ field of nearly 5 nV/cm, which at a horizontal component of the earth's magnetic field of 20 μT corresponds to a swimming velocity of 2 cm/s, conservatively rated.

Since the afferent electrophysiological nerve response is approximately linear from the threshold value to eight times the threshold value, the dynamic range of the individual longest canal ampullae is ≈ 5–40 nV/cm. Nevertheless, the dynamic range of the ampullary system is much larger, since the pore-to-pore distances of differential ampullae vary from very small to great lengths. Below the narrow, eight-fold dynamic range of the individual ampullary sense organs, the signal is swamped by the system noise. Above this range, the individual ampullae quickly saturate, as judged from the afferent action potential rates. Actually, the receptor membranes are active elements, as assumed in this above paragraph. Amplification of the electrical signal resulting from the negative receptor current causing positive feedback will be discussed under Sect. 2.3.2.

2.2.3 *Further Developments: Dependence on Resistivity of Medium*

2.2.3.1 Effect of Loading of Animal by Return Current Through Surrounding Medium in Review

In the electrically heavily loaded, in the open ocean traveling shark, the charge accumulations at opposite skin surfaces are rather small. Hence, the resulting $\nabla\varphi_{\mathrm{anl}}$ is rather weak, offering the animal in ocean waters a $\int\nabla\varphi_{\mathrm{mem}} \cdot \mathbf{ds} = \int -\rho\mathbf{J}_{\mathrm{anl}} \cdot \mathbf{ds}$ in strength close to, though not equal to the $-\int \mathbf{v} \times \mathbf{B}_{\mathrm{h}} \cdot \mathbf{ds}$. However, in a shark entering or permanently living in fresh waters, thus getting or being less heavily loaded, the $\nabla\varphi_{\mathrm{anl}}$ is not insignificant, and the $\int\nabla\varphi_{\mathrm{mem}} \cdot \mathbf{ds}$ differs in strength from the $-\int \mathbf{v} \times \mathbf{B}_{\mathrm{h}} \cdot \mathbf{ds}$ by $\int\nabla\varphi_{\mathrm{anl}} \cdot \mathbf{ds}$, rendering the system less effective. In the extreme case of a bird flying in dry air at a constant velocity of 100 km/h, the strong $\int\mathbf{v} \times \mathbf{B}_{\mathrm{h}} \cdot \mathbf{ds}$ it induces is all but completely offset by the nearly equally strong $\int\nabla\varphi_{\mathrm{anl}} \cdot \mathbf{ds}$ of the charges accumulated on the skin on opposite sides of the animal, rendering $\int\nabla\varphi_{\mathrm{mem}} \cdot \mathbf{ds}$ practically zero, showing dramatically that it is the $-\rho\mathbf{J}_{\mathrm{anl}}$, not the $\mathbf{v} \times \mathbf{B}_{\mathrm{h}}$, that measures the physically adequate stimulus. True, in flying birds, relative motions between body parts due to the flapping of the wings do result in motionally induced electric currents, though the use of such fluctuating signals has not found physiological support thus far.

2.2.3.2 Critical Importance of Charge Accumulations

The surface charges on the skin of the animal are largely those due to less than complete loading, diminishing the $-\rho\mathbf{J}_{\mathrm{anl}}$. Yet, there are also very small enhancing surface charges just outside the skin pores, building up to direct the electric current across the body to flow around but not through the ampullary sense organs. Here, we will neglect these small enhancing charges and consider the charges on the skin only to be those degrading the physically adequate stimulus.

2.2.3.3 To Summarize

Let us consider three levels of electrically loading a traveling shark by the return current through surrounding medium.

A traveling shark receives the pore-to-pore $\int -\rho\mathbf{J}_{\mathrm{anl}} \cdot \mathbf{ds}$, which equals $\int(\mathbf{v} \times \mathbf{B}_{\mathrm{h}} - \nabla\varphi_{\mathrm{anl}}) \cdot \mathbf{ds}$.

Upper Limit of Heavily Loaded Shark No $\nabla\varphi_{\mathrm{anl}}$ by lack of charges accumulations on opposite skin surfaces. The energy imparting $\int\mathbf{v} \times \mathbf{B}_{\mathrm{h}} \cdot \mathbf{ds}$ in the tissues is completely offset by the energy dissipating $\int -\rho\mathbf{J}_{\mathrm{anl}} \cdot \mathbf{ds}$. Hence, the physically adequate stimulus $\int -\rho\mathbf{J}_{\mathrm{anl}} \cdot \mathbf{ds}$ equals $-\int \mathbf{v} \times \mathbf{B}_{\mathrm{h}} \cdot \mathbf{ds}$.

Lower Limit of Hardly Loaded Animal No $-\rho\mathbf{J}_{\mathrm{anl}}$ by lack of $\mathbf{J}_{\mathrm{anl}}$ across the body. The energy $\int\mathbf{v} \times \mathbf{B}_{\mathrm{h}} \cdot \mathbf{ds}$ imparted to the tissues is completely stored as the energy $\int\nabla\varphi_{\mathrm{anl}} \cdot \mathbf{ds}$ of the field of the charge accumulations. Hence, the physically adequate stimulus $\int -\rho\mathbf{J}_{\mathrm{anl}} \cdot \mathbf{ds}$ is nil.

Moderately Loaded Animal $\int\mathbf{v} \times \mathbf{B}_{\mathrm{h}} \cdot \mathbf{ds}$ in the tissues is partially balanced by $\int -\rho\mathbf{J}_{\mathrm{anl}} \cdot \mathbf{ds}$ and partially adds to $\int\nabla\varphi_{\mathrm{anl}} \cdot \mathbf{ds}$. Hence, the physically adequate stimulus $\int -\rho\mathbf{J}_{\mathrm{anl}} \cdot \mathbf{ds}$ equals a fraction of $-\int \mathbf{v} \times \mathbf{B}_{\mathrm{h}} \cdot \mathbf{ds}$.

Sharks in ocean waters are most well loaded, so that their $\int -\rho\mathbf{J}_{\mathrm{anl}} \cdot \mathbf{ds}$ is closest to $-\int \mathbf{v} \times \mathbf{B}_{\mathrm{h}} \cdot \mathbf{ds}$. However, in freshwater sharks lose their superior efficiency in detecting motional electric fields.

2.2.3.4 Pared-Down Version of Comprehensive Electric and Magnetic Orientation Theory

At this stage, it may be opportune, though still treacherous, to restate the conceptually more user-friendly version of the author's formal, comprehensive theory concerning the shark's detection of its magnetic compass heading. Be aware however, the pared down version has serious caveats: (1) it is a reasonable approximation *only* under strict, limiting conditions, and (2) scholarly speaking, it is *incorrect*, as you may have gleaned from the thorough mindboggling analysis conducted thus far. Keeping these warnings in mind, we may proceed, although with trepidation, because, outside its range of reasonable applicability, it leads to erroneous conclusions. Let us review the issues briefly and look at them from different sides. They already have led too many scientists astray.

A shark moving with velocity $\mathbf{v}$ in ocean waters is, according to an earth-bound observer—by interaction with the horizontal component of the earth's magnetic field $\mathbf{B}_{\mathrm{h}}$—subject to a ventro-dorsal motionally induced electric field $\mathbf{v} \times \mathbf{B}_{\mathrm{h}}$, which along a pore-to-pore path through the ampullary canals increases the voltage by $\int\mathbf{v} \times \mathbf{B}_{\mathrm{h}} \cdot \mathbf{ds}$. Along the same path, the voltage across the sensory epithelium in the blind ampulla proper increases by $\int\nabla\varphi \cdot \mathbf{ds}$. What is more, along a pore-to-pore path through the skin and body tissues, bypassing the ampullary sense organs, the $\mathbf{v} \times \mathbf{B}_{\mathrm{h}}$

increases the voltage by the same $\int \mathbf{v} \times \mathbf{B}_h \cdot \mathbf{ds}$ across the skin and body tissues. Hence, in any measuring circuit through the ampullary sense organs closing through the skin and body tissues, the $\int \mathbf{v} \times \mathbf{B}_h \cdot \mathbf{ds}$ drops out of the equation. Furthermore, any electrostatic field between opposing skin pores will also result in the same voltage increase along the ampullary path and the skin-body tissue path, hence dropping out of the equation as well. Since in the ampullary path, the sensory epithelium is an excellent insulator, no electric current flows through the sense organs, but the electric current flows freely along the path through the skin and body tissues, producing a voltage increase $\int - \rho\mathbf{J}_{anl} \cdot \mathbf{ds}$ along the latter path. Hence, around the measuring circuit, the remaining voltage increase $\int \nabla\varphi \cdot \mathbf{ds}$ along the ampullary path, to wit, across the sensory epithelium, equals the remaining voltage increase $\int - \rho\mathbf{J}_{anl} \cdot \mathbf{ds}$ along the path through the skin and body tissues. Ergo: according to the earth-bound observer, the shark detects the $\int - \rho\mathbf{J}_{anl} \cdot \mathbf{ds}$, being oblivious of the earth-bound observer's $\int \mathbf{v} \times \mathbf{B}_h \cdot \mathbf{ds}$.

Well, we need not be surprised, since the shark does not move with a velocity $\mathbf{v}$ in its proper frame of reference in which it makes the measurements. In the animal's frame, there is no $\mathbf{v} \times \mathbf{B}_h$ in either the ampullary canals or in the skin and body tissues. The animal does not "know" of a magnetic field in which it moves, but according to Einstein it receives a particular ambient electric field instead. Actually, we ourselves in the earth frame may think in the animal's frame the environment to move in the earth's magnetic field, thereby inducing an electric field invading the animal. But how does the $\mathbf{v} \times \mathbf{B}_h$ driving the induced electric current resulting in the $-\rho\mathbf{J}$ relate to the $-\rho\mathbf{J}$? This relationship is given by the fundamental equation $\nabla\varphi = \mathbf{v} \times \mathbf{B} - \rho\mathbf{J}$. Hence, the difference is the voltage gradient due to the electrical charges accumulating on opposite skin surfaces of the animal. Ergo: the fundamental difference between the $\mathbf{v} \times \mathbf{B}_h$ driving the induced electric current resulting in the $-\rho\mathbf{J}_{anl}$ is given by the electric field of the charge accumulations on the skin of the animal.

Albeit, since in ocean waters sharks are heavily loaded and, hence, the charge buildup on the skin is minimal, the pore-to-pore $- \int \mathbf{v} \times \mathbf{B}_h \cdot \mathbf{ds}$ through the ampullary canals and capsules offers the observer at rest in the earth frame a fairly good approximation of the pore-to-pore $- \int \mathbf{v} \times \mathbf{B}_h \cdot \mathbf{ds}$ the shark detects. Hence, an observer at rest in the earth frame may consider $\mathbf{v} \times \mathbf{B}_h$, in the ampullary canals and the ampullary capsules as building up the charges across the sensory epithelia, the electric field of which causes the excitation of the electroreceptive voltage-sensitive ion channels. Yet, it is merely an approximation and only applicable to animals that are heavily loaded by the ionic media with respect to which they are moving and, hence, only endure a small built-up of charges on the skin that reduce the physically adequate stimulus. Shorted animals are a fiction. To repeat, the animal detects the $\int - \rho\mathbf{J}_{anl} \cdot \mathbf{ds}$, whether considered in the earth frame or in the animal frame, whereas the earth-bound observer says, well, that is in electrically heavily loaded sharks almost the same as the $- \int \mathbf{v} \times \mathbf{B}_h \cdot \mathbf{ds}$ as we see it in our earth frame. Since the $- \int \mathbf{v} \times \mathbf{B}_h \cdot \mathbf{ds}$ offers about the same information as the $\int - \rho\mathbf{J}_{anl} \cdot \mathbf{ds}$ and expresses that information in terms we understand much better, why should we not use the $-\int \mathbf{v} \times \mathbf{B}_h \cdot \mathbf{ds}$ instead of the $\int -\rho \mathbf{J}_{anl} \cdot \mathbf{ds}$? Alright, granted if it is understood that it is only an approximation and is only defensibly valid for sharks heavily loaded by the seawater medium. Yet, $-\int \mathbf{v} \times \mathbf{B}_h \cdot \mathbf{ds}$ is such a handy "proxy" for $\int -\rho \mathbf{J}_{anl} \cdot \mathbf{ds}$ that I will also use it in the next section.

True, when two bare electrodes, that is, electrodes *not* embedded in a moving body, are towed through still seawater in the presence of the earth's magnetic field, the motional electric signal a voltmeter moving along with the electrodes receives results only from the electric field of the charges building up *across the voltmeter's sensing element* to balance the $\mathbf{v} \times \mathbf{B}$ in the salt bridge to voltmeter leads. This simple situation of towed electrodes does not apply to a swimming shark, though, as the induced currents in the shark's body tissues build up charges at opposite skin surfaces of the animal, the field of which opposes the $\mathbf{v} \times \mathbf{B}$ in the ampullary canals, hence reduces the electrical charges across the sensory epithelium. Though the charges on opposite skin surfaces are reduced by the return current through the still seawater medium, with respect to which the animal moves, they can never be completely eliminated, as they are needed to drive the return current though the seawater medium for the motionally induced current in the skin and body tissues and the seawater has a finite resistance.

Let us review this important issue once more. To repeat, the $\mathbf{v} \times \mathbf{B}$ in the animal's skin and body tissues drives an electric current across the body. This current causes a surface charge buildup on the skin on opposite sides of the body, which, in turn, causes a return current though the seawater medium with respect to which the animal moves. Yet, at the same time the electric field of the *charge buildup on the skin* counteracts the $\mathbf{v} \times \mathbf{B}$ in the animal's skin and body tissues and in the ampullary canals at the cost of the excitation *charges on either side of the sensory epithelium.* The skin charges will be small if the return current carries them off quickly, that is, if the return path through the seawater heavily loads the shark. However, if the return current is only small, that is, if the return path through the outside medium, freshwater or thin air, is higher ohmic, the charge buildup may be high, in the limit causing an electric field almost completely offsetting the $\mathbf{v} \times \mathbf{B}$, leaving hardly anything for the *charges across the sensory epithelia.* In short, oceanic sharks are blessed by having a body plus skin resistance some hundred times larger than the seawater they replace.

2.2.3.5 Effect of Vertical Component of Earth's Magnetic Field

So far, interaction of the traveling shark with the vertical component of the earth's magnetic field $\mathbf{B}_v$ has not yet been discussed. For a shark well under the water surface, the situation is basically the same as that resulting from interaction with the horizontal component of the earth's magnetic field $\mathbf{B}_h$ except that the whole picture is symmetrical about a horizontal, rather than the vertical axis, that is, ninety degrees rotated. However, the biological significance of interaction with vertical component of the earth's magnetic field is very different from the interaction with the horizontal component giving the animal its compass heading, as will below be explained how.

2.2.3.6 Comparison of Electric Fields from Ocean Currents and from Animal's Own Swimming Movements

It is worth emphasizing that the underlying physics of the drifting shark detecting its drift and the traveling shark detecting its magnetic compass heading are the same. In the instance of the drifting shark, it is the stream, containing the shark, that is electrically loaded by deeper water and the ocean bottom, whereas in the instance of the traveling shark, it is the shark, containing the electroreceptive organs, that is electrically loaded by the water relative to which it is moving. However, if we would apply the pared-down version of the comprehensive theory to the drifting shark, we would in general go badly wrong. The pared-down version would apply only to short-circuited streams with negligible space charge accumulations at their boundaries, not to lightly loaded streams as those extending to the bottom, leaving only the ocean floor as the return path, and certainly not to a tidal stream in a fjord, where the electric field of the charge accumulations virtually completely offsets the motionally induced electric field, rendering the $-\rho\mathbf{J}$ near nil. Hence, the shortcut version of the electric and magnetic orientation theory can indeed be very misleading, when it is not thoroughly understood that its approximate validity is strictly limited to electrically heavily loaded ocean streams and sharks.

2.2.3.7 Strength of Motional Electric Fields

To give an idea of the magnitudes of the $-\rho\mathbf{J}$ the animal receives, let the animal move normal to the magnetic field. Let the horizontal component of the earth's magnetic field be 25 μT. Then at a cruising speed of 50 cm/s, the stimulus the animal receives is 25 nV/cm, ample for the animal to determine its compass heading.

2.2.4 Relaxing of Magnetic Constraints, Extending Models to Meet Nature

As promised, the simplifying assumptions made to explain the shark's motional electric detection of the earth's magnetic field would be relaxed so as to bridge the gap between the model shark and the real shark. Electrically, the greatest variations result from the degree of loading and from the shape of the animal. The loading changes drastically when a shark enters fresh waters, undergoing an increase in resistivity from 20 Ω cm to 2 kΩ cm or more, rendering the motional electric detection of the earth's magnetic field less and less plausible. As for the shape of the animal, if it were ellipsoidal, would have homogeneous resistivity, and the $\mathbf{v} \times \mathbf{B}_h$ were uniform, the $\mathbf{E}_{anl}$ and $-\rho\mathbf{J}_{anl}$ would in the animal be uniform as well (Fig. 2.5). If the shape of the animal is of any other form, but the $\mathbf{v} \times \mathbf{B}_h$ still remains uniform, the $\mathbf{E}_{anl}$ and $-\rho\mathbf{J}_{anl}$ get curved, though such as to keep their vector sum uniform in conformity with the $\mathbf{v} \times \mathbf{B}_h$, which they are in dynamic equilibrium with. The thinner the animal is dorsoventrally, as in the case of a ray, the less sensitive it is to the horizontal component of the earth's magnetic field due to the large charge built-up on the dorsal and ventral skin surfaces and the shorter pore-to-pore distances, and the more sensitive it is to the horizontal component of the ambient electric field due to the small charge built-up and the longer pore-to-pore distances. Might the undulatory movements of the large pectoral fins of rays make up for the lesser efficiency of detecting the horizontal component of the earth's magnetic field? See below.

Changes in the swimming speed cause, by interaction with the horizontal component of the earth's magnetic field, changes in the $\mathbf{v} \times \mathbf{B}_h$, the charge accumulations at the animal's ventral and dorsal skin surfaces, and the vorticity at its left and right sides, such as proportionally to alter the strength of the vertical $-\rho\mathbf{J}_{anl}$ the animal detects, and by interaction with the vertical component of the earth's magnetic field, changes in the swimming speed cause changes in the $\mathbf{v} \times \mathbf{B}_v$, the charge accumulations at the left and right skin surfaces, and the vorticity at the ventral and dorsal skin sides, such as proportionally to alter the strength of the horizontal $-\rho\mathbf{J}_{anl}$ the animal detects.

Changes in the direction of swimming, without changes in swimming speed, do not cause changes in the $\mathbf{v} \times \mathbf{B}_v$, but lead to changes in the $\mathbf{v} \times \mathbf{B}_h$, the charge accumulations at the ventral and dorsal skin surfaces, and the vorticity at its left and right skin surfaces, such as to alter the strength of the vertical $-\rho\mathbf{J}_{anl}$ field the animal detects. During the turns, though, the animal induces also vorticity, that is, loop currents flowing entirely within the body. Interestingly, the design of the ampullae of Lorenzini with the skin pores opening to the outside seawater is such that the animal is not effectively stimulated by those internal loop currents. Changes in the shape of the animal, such as during the sharks' meandering mode of swimming and the undulating wing flapping of rays also cause induced electric currents to flow between the different body parts of the animal.

As for the body of animal, the inherent non-homogeneous resistivity of the skin and tissues might be thought to complicate the situation dramatically, since the $-\rho\mathbf{J}_{anl}$ depends inherently on the resistivity ρ. However, when we say that the animal detects the $-\rho\mathbf{J}_{anl}$, we mean, it detects the pore-to-pore line integral of $-\rho\mathbf{J}_{anl}$ across the body. And, since at uniform $\mathbf{v} \times \mathbf{B}$, the $-\rho\mathbf{J}_{anl}$ is conservative, this pore-to-pore line integral is a definite quantity, however complicated $-\rho\mathbf{J}_{anl}$ may be inside the animal, see Appendix 4. Hence, the problem is solved by the integrative properties of the system, evening out the $-\rho\mathbf{J}_{anl}$ irregularities along the path of integration.

So far, I have dodged one fundamental issue. In general, sharks do not follow straight paths at constant velocity for any length of time. If they do so anyhow, their electroreceptors will adapt to the DC ambient electric field and to the DC electric field resulting from their motion in the earth's magnetic field within seconds. Hence, the animals must vary their velocity in order to modulate the ambient-electric and motional-electric fields and bring them into the low-frequency range of the sense organs. How then shall they change their velocity to detect the DC fields resulting from their drift in ocean currents or from swimming in the earth's magnetic field? They either may change their speed without changing direction, or they may change the direction without changing their speed or change both their speed and the direction. These topics will be discussed in the next section, dealing with the animals' exploratory movements.

2.2.5 *Biological Interpretation: Detection of Compass Heading*

2.2.5.1 Compass Heading and Magnetic Latitude

What may the traveling shark learn from the pore-to-pore $\int -\rho\mathbf{J}_{\text{anl}} \cdot \mathbf{ds}$ it detects in its own frame of reference? By making the proper exploratory movements, the animal may gauge, what the stationary observer calls, the $-\int \mathbf{v} \times \mathbf{B}_{\text{h}} \cdot \mathbf{ds}$, corrected for the charge accumulations on the skin due to the incomplete loading, and thereby the strength and direction of $\mathbf{B}_{\text{h}}$, the horizontal component of the earth's magnetic field, hence its magnetic compass heading. Of course, the shark does not and cannot infer the $-\int \mathbf{v} \times \mathbf{B}_{\text{h}} \cdot \mathbf{ds}$ reckoned by an earth-bound observer, neither does it have a knowledge of the charge accumulations on the skin. Nevertheless, it seems to behave as if it does, by instinctively orienting appropriately to the $-\rho\mathbf{J}_{\text{anl}}$ fields it receives, based on innate and gained experience. Our reasoning merely proves that the necessary information is available, whether viewed in the earth frame, where we may speak of a *magnetic* compass sense, or in the shark frame, where we may speak of a *motional electric* compass sense. In this respect, sharks are more accomplished than we are in practicing textbook physics. We have to reason it through to overcome our man-made intricacies. For sharks, it is second nature. Yet, now we know, it is not difficult to program the author's shark-emulating robotic device to set its compass heading by detecting the electric fields resulting from its properly executed exploratory movements.

By also detecting the vertical component $\mathbf{B}_{\text{v}}$ of the earth's magnetic field—as an earth-bound observer calls it—the shark may furthermore gauge the local inclination of the earth's magnetic field, hence its magnetic latitude on the earth's globe. Moreover, interaction with the vertical magnetic component $\mathbf{B}_{\text{v}}$ may give the animal a measure of its speed of swimming independent of its compass heading. For further detail, see Sect. 2.2 of the present article.

2.2.5.2 Drift in Terms of Swimming Speed, a Hypothetical Scenario

Above, the question was raised how a shark may gauge the velocity of its drift with respect to still or slower moving deeper water from the $-\rho\mathbf{J}_{\text{str}}$ it receives in an open-ocean stream, hence, how to calibrate the $-\rho\mathbf{J}_{\text{str}}$ in terms of the stream's velocity. In the following worked-out example, we will assume the animal to be aware—from being far out in open-ocean waters or from having verified by exploration—to be in a nearly shorted wind-driven ocean stream. The objective here is to show that it is logically possible for the animal to assess its drift velocity from the $-\rho\mathbf{J}_{str}$, even though it may not do so as outlined below. Yet, I am convinced that, if we can do it our way, sharks will certainly be able to do it their way, probably much more elegantly.

Let us first explore the technical possibility of assessing the velocity of a virtually shorted ocean stream, interacting with the vertical component of the earth's magnetic field, by operating from a raft equipped with two submersed horizontally directed salt bridges and electrodes connected to a low-frequency AC voltmeter. While drifting, let us align the electrodes with the direction of the stream, that is, normal to the direction of $-\rho\mathbf{J}_{str}$, for a time long enough to reset the voltmeter to null. Then, let us turn the raft ninety degrees to have it face upstream so as to orient the electrodes normal to the stream to give us the value of $-\rho\mathbf{J}_{str}$ and, in a time short compared to the adaptation time of the voltmeter, tow the electrodes by the raft upstream at increasing velocity until the voltmeter reads null again. At that moment, our speed with respect to the water is a measure of the speed of the ocean stream with respect to deeper water, though in opposite direction, assuming the charge accumulations at the boundaries of the heavily loaded stream indeed to be negligible. From the theory presented above, the reader should be able to follow my reasoning and agree, or disagree, with my conclusion.

This scenario, though hypothetical, may serve to hone our skills in exploring the animal's sensory world. However, for the animal to execute our technical procedure, it has to take the electrical properties of its body into account, as it detects not only the $-\rho\mathbf{J}_{str}$ but also the $-\rho\mathbf{J}_{anl}$. I will leave that as an exercise for the animals. Note, furthermore, that the signal the animal receives from its active interaction with the earth's magnetic field is proportional to its speed with respect to the flowing water, not with respect to still deeper water or the earth. The described scenario relies on the horizontal electric field induced by interaction with the vertical component of the earth's magnetic field. By also recording the vertical electric field by interaction with the horizontal component of the earth's magnetic field with two vertically directed electrodes, the animal also gets information about the magnetic direction of the stream.

2.2.6 Compass Heading Sensed in Frame of Shark

As discussed for the shark drifting in an ocean stream, again the question arises what the shark detects in its own frame of reference in which neither the animal nor its ampullary sense organs are moving. The earth's magnetic field $\mathbf{B}$ and the motionally induced electric field $\mathbf{v} \times \mathbf{B}$, envisioned by the earth-bound observer, have for the animal no sensible meaning. Those terms are merely human constructs. Still, the results in the animal's rest frame are consistent with those of the observer at rest in the earth frame, since classical theory and the theory of relativity once more conclude, that a shark by moving in a *magnetic* field in the earth frame, is exposed to an ambient *electric* field $\mathrm{E}'_{\mathrm{amb}}$ in its own rest frame. The ambient electric field $\mathrm{E}'_{\mathrm{amb}}$ in the shark frame nulls the $\mathbf{v}' \times \mathbf{B}'$ of the whole moving part of the world, and penetrates the parts that are at rest, here only the shark itself, in which it gives rise to the $-\rho\mathbf{J}'_{\mathrm{anl}}$ the animal detects. The primes of the **B** and **J** will be omitted, since their relativistic differences in the two frames of reference are negligible, so that in both the earth frame and the shark frame the physically adequate stimulus $-\rho\mathbf{J}_{\mathrm{anl}}$ is, at the limiting speeds considered, the same. Hence, the shark's sensory world is purely electric, not magnetic, even though we accredit the animals with a *magnetic* compass sense, simply because we are so self-centered as to study sharks from our point of view, not theirs. For the animals it is an *electric* compass sense, as I learned from Einstein (1905). Remember, though, the imposed electric field of an ocean stream and the electric field induced by the animal itself are totally different functions of the animal's exploratory movements.

2.3 Additional Remarks

2.3.1 Relative Motions Between Body Parts

2.3.1.1 Turning Motions

Until now, we have only considered the rectilinear, *whole-body* motions of sharks and rays traveling in different compass directions, with emphasis on their interaction with the horizontal component of the earth's magnetic field. All induced electric currents traversed the body, left the animal through the skin on the one side, and re-entered through the skin on the other side, thus closing their loops through the stationary environment. Most of the stimulus voltage thereby collected occurred by the current crossing the resistive skin. However, when the animal interacts with the horizontal component of the earth's magnetic field by turning on the spot, that is, by merely rotating about its dorso-ventral body axis, the motionally induced electric current flow ventro-dorsally on the one side of the body and dorso-ventrally on the other side, leading to current loops almost entirely remaining under the skin, which offers much higher resistance than the body fluids, hence without significantly entering into the surrounding seawater, and thus without effectively stimulating the ampullary sense organs, which rely on the external potentials at the skin pores.

2.3.2 *Reactance of Tissues and Excitability of Receptor-Cell Membranes*

2.3.2.1 Reactance of Body Tissues

So far, we have limited the discussion to purely resistive models under stationary conditions. However, if we take the capacity of the wall of the ampullary canal and the ampulla proper into account, the integrating time constant of the electrical signal reaching the receptor epithelium is for long canals on the order of 50 ms (Waltman 1966). This will not significantly infringe on the low-frequency range of the signals the animals use to detect their drift and compass headings. Furthermore, the adaptation to DC stimuli taking place in the sense organs is sufficiently slow, with a time constant on the order of seconds, for the animal to have time to execute its fast exploratory moves within the time frame in which the adaptation is still of little concern.

2.3.2.2 Negative Resistance and Positive Feedback

Furthermore, we have left out the excitability of the receptor-cell membranes, which is a much livelier issue. A positive voltage received at the skin pore, relative to the basal stroma side of the ampulla, spreads along the gel-filled canal into the ampulla proper, causing the stimulus voltage to concentrate across the sensory epithelia. In response, the ion channels in the apical membranes of the receptor cells produce a negative signal current, flowing from the ampullae to the pores, which makes the voltage in the canal and ampulla proper from the skin pore to the ampulla to incline, instead of decline, thus increasing the electrical signal. Moreover, the negative current causes, by interaction with the basal membranes of the receptor cells, the voltage inside the receptor cells to swing far negative, many times the amplitude of the positive stimulus applied, thereby hyperpolarizing the basal membranes and causing, after synaptic transmission, a decrease in the action-potential rate in the afferent ampullary nerves. Similarly, a negative voltage received at the skin pore causes a positive swing inside the receptor cells, resulting in an increase in the action-potential rate. Hence, aside from the fact that the stimulus voltage received across the animal is passively focused on the receptor-cell epithelia, greatly increasing the strength of the signal, the receptor-cell epithelia themselves amplify the stimulus actively several fold by the process of positive feedback (see Fig. 2.8).

2.3.2.3 Frequency Range in Terms of Signal to Noise

The author has discovered a curious, as of yet not fully understood phenomenon in processing the action-potential trains in the afferent nerves of the ampullae and of the lateral-line and inner-ear sense organs. When he decodes the nerve trains by first phase-binning the action potentials, then least-square fitting the results to the applied

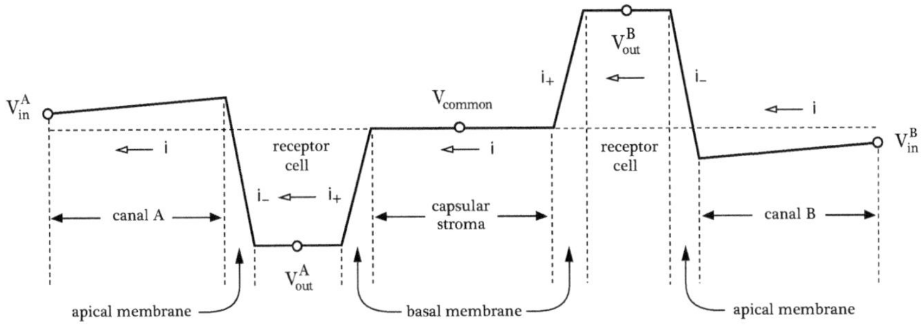

Fig. 2.8 Voltage and current distribution in differential pair of electroreceptive ampullae in sharks and rays. V_{in} voltages at skin pores, V_{out} voltages across basal membranes, V_{common} voltage in capsule, *i* negative input current, i_- current through *negative* resistance apical membrane i_+ current through *positive* resistance basal membrane. Due to positive feedback amplified voltages across basal membranes lead to synaptic transmission and subsequently to action-potential generation in afferent nerve fibers

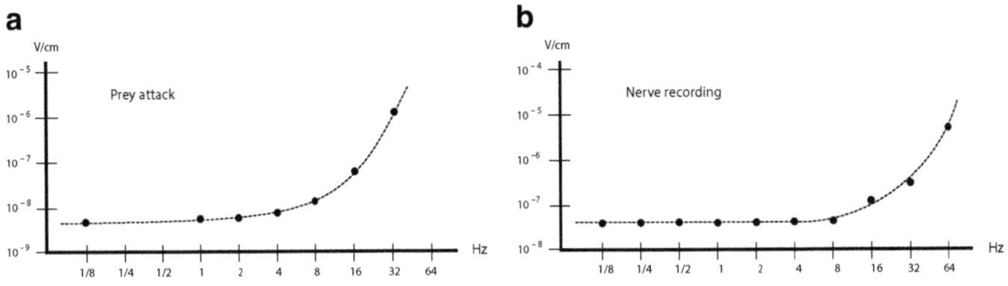

Fig. 2.9 Comparison of behavioral and electrophysiological frequency response. (**a**) Behavioral feeding responses of thornback ray, *Raja clavata*, directed at current source of prey-simulating dipole field of wide-ranging sinewave frequencies. Average of year-long observations. Response threshold remained practically constant through seasons despite very different, although slowly changing temperatures, from near freezing to undue hot in Dutch WW1 underground bunker. (**b**) Single fiber recording from ampullary nerve in thornback ray, *Platyrhinoidis triseriata*, in terms of action-potential rate in response to uniform electric field applied to submerged animal. This is the signal-to-noise threshold vs. frequency, *not* a conventional amplitude threshold curve that contradicts the extreme low-frequency sensitivity seen behaviorally

stimulus, and in the end applying an analysis of variance to the results, the outcome completely suppresses the inherent 1/*f* noise of the system, rendering the sense organs sensitive to near-DC, just as the behavioral experiments show (Fig. 2.9).

Presently, it seems that the high electrical sensitivity of the ampullary sense organs is not so much due to extremely low-noise electrical-to-electrophysiological transducers, but rather to a very clever processing of the nerve signals. Most intriguing is the possible function of the lone kinocilium protruding from the apical surface of the receptor cells into the gelatinous contents of the ampullae proper. Evidently, further biophysical and electrophysiological studies will be needed to determine the role of the kinocilium, supposedly featuring the voltage-sensitive ion channels, more about the biophysical operation of the transducers, and the special kind of nerve-action-potential coding, decoding, and the central processing.

2.3.3 *Passive and Active Electro-orientation: Concluding Remarks*

2.3.3.1 Relativistic Approach

A most fascinating issue is the parallel between a shark moving in the earth's magnetic field, that is, moving with respect to the stationary considered source of the earth's magnetic field, and Einstein's copper coil moving with respect to a stationary considered magnet or, vice versa, the source of the earth's magnetic field moving with respect to the stationary considered shark and Einstein's magnet moving with respect to a stationary considered copper coil, by which he introduced his original 1905 article on the special theory of relativity. For a reader familiar with special relativity, the present discourse concerning the animals' view of the world must be self-evident in that the answers to the questions posed as to the transformation of electric and magnetic fields from the familiar earth frame to the shark frame of reference for them are well known in advance from Einstein's reasoning. In short, the animals remind us that Einstein's theory of special relativity is relevant even at speeds as low as a shark cruising ocean waters. In that respect, Faraday's empirical law was only a steppingstone toward Einstein's deeper understanding of the physics "as it was understood at his time." To appreciate Einstein's 1905 article, one should also consult Föppl's 1904 classical German textbook, from which Einstein learned the theory of electricity.

2.3.3.2 Is It Just the Lorentz Force?

An elementary question is often raised: is it not simply the Lorentz force, $\mathbf{L} = q(\mathbf{E} + \mathbf{v} \times \mathbf{B})$, of the author's distinguished Dutch compatriot, that excites the shark's electroreceptors by acting on the gating charges of the voltage-sensitive ion channels? The answer is a hesitating "yes of course," but—perhaps surprisingly—it is not so much the motionally induced $\mathbf{v} \times \mathbf{B}$, but rather the electrostatic $\mathbf{E}_{\mathrm{mem}}$ of the Lorentz force that acts on the gating charges of the ion channels. The $\mathbf{v} \times \mathbf{B}$ part of the Lorentz force acting on the gating charges actually is negligibly small. Certainly, in ocean streams and in traveling sharks, the $\mathbf{E}_{\mathrm{mem}}$ field derives from the $\mathbf{v} \times \mathbf{B}$ field across the animal, yet the line integral of $\mathbf{E}_{\mathrm{mem}}$ across the receptor membrane may range from near zero in sharks drifting in unloaded streams and birds flying in thin air to even more than the full value of the line integral of $\mathbf{v} \times \mathbf{B}$ across the animal in sharks drifting in nearly shorted streams. It is the resulting pore-to-pore integral of $-\rho\mathbf{J}$ that measures what the animal detects. How effectively the driving $\mathbf{v} \times \mathbf{B}$ field leads to the excitation forces on the gating charges of the voltage-sensitive ion channels critically depends on the charge accumulations at the boundaries of the stream, on the skin of the sharks, and in the ampullary sense organs. Those, in turn, depend on the resistivity distribution of the various parts of the current circuit. The details are overwhelming and will be left for my successors.

Another subtle issue is at stake concerning the Lorentz force, as it is the total *electrical force* on the ions, that is the electric force proportional to the charge of the ions. However, *non-electrical forces* play a crucial role as well. In particular, the ohmic drag force $-\rho\mathbf{J}$, which is a frictional force, not proportional to the charge of the ions, though the resistivity $-\rho\mathbf{J}$ is defined such as to make the drag force proportional to the charge of the ions. Hence, $-\rho\mathbf{J}$ is not included in the Lorentz force, nor in Maxwell's free-space equations. Nevertheless, it is the physically adequate stimulus in drifting as well as in swimming sharks and rays. True, it is the motionally induced electric $\mathbf{v} \times \mathbf{B}$ and the ambient electric field $\mathbf{E}$ that cause the various forces, of which the $-\rho\mathbf{J}$ is what the sharks and rays detect in the sense that it is the $-\rho\mathbf{J}$ that measures the physically adequate stimulus. Do not forget, in the frame of the shark—who does the sensing—its velocity $\mathbf{v}$ is zero, yet the $-\rho\mathbf{J}$ is practically the same as the $-\rho\mathbf{J}$ the earth-bound observer arrives at. In that respect transformation of the physically adequate stimulus between the earth and shark frames of reference is very simple.

2.3.3.3 Exploratory Movements

It is reassuring, at least for us, when a shark actively probes the earth's magnetic field and deliberately changes the speed and direction of its velocity, it may be considered gradually to move through a sequence of quasi-steady-state situations, except for the effects of their sense organs physiologically having finite times of integration and differentiation. Hence, practically speaking, the stimulus fields are all but quasi-static.

I repeat for emphasis, many intrigued readers have been bewildered by the question as to how sharks discriminate between the electric fields informing them about their drift with ocean streams and the electric fields informing them about their magnetic compass headings. The resolution rests upon the observation that the two kinds of electric fields sharks receive, about their drift and about their compass heading, are fundamentally different functions of the animals' motions. The question as to how sharks direct their seemingly erratic swimming movements so as to collect the different kinds of sensory data will be the topic of next Sect. 2.2 of the present article.

Appendices

Appendix 1. Definition of Symbols in Summary Statements

Summary Statement 1 for Wind-Driven, Open-Ocean Stream

${}^{str}\mathbf{J}_{str}$ denotes the induced electric current of density $\mathbf{J}$, by superscript considered in the stream, by subscript driven by the $\mathbf{v} \times \mathbf{B}$ in the stream, yet impeded by the resistive drag force, and countered by the motional electric field of the electrical space charges accumulated at the lateral boundaries of the stream.

$-\rho^{str}\mathbf{J}_{str}$ denotes the resistive force $-\rho\mathbf{J}$ on the ionic charge carriers, by superscript considered in the stream with ρ the resistivity in the stream, by subscript resulting from the electric current of density $^{str}\mathbf{J}_{str}$ motionally induced in the stream. The $-\rho^{str}\mathbf{J}_{str}$ is the field drifting electrodes detect.

$^{str}\mathbf{E}_{str}$ denotes the motional electric field $\mathbf{E}$, by superscript considered in the stream, by subscript due to the electrical space charges accumulated at the lateral boundaries of the stream.

$^{stw}\mathbf{J}_{str}$ denotes the electric current of density $\mathbf{J}$, by superscript considered in still water under and adjacent to the stream, by subscript driven by the space charges accumulated at the lateral boundaries of the stream.

Summary Statement 2 for Shark Drifting in a Wind-Driven, Open-Ocean Stream

Hint: Since $\mathbf{v} \times \mathbf{B} - \rho\mathbf{J} + \mathbf{E} = 0$, at constant $\mathbf{v} \times \mathbf{B}$ the resistive force $-\rho\mathbf{J}$ increases when the motional electric field $\mathbf{E}$ decreases.

In the stream:

$^{str}\mathbf{J}_{str}$ denotes the induced electric current of density $\mathbf{J}$, by superscript considered in the stream, by subscript driven by the $\mathbf{v} \times \mathbf{B}$ in the stream, though impeded by the resistive force in the stream, and countered by the motional electric field of the electrical space charges accumulated at the lateral boundaries of the stream.

$-\rho^{str}\mathbf{J}_{str}$ denotes the resistive force $-\rho\mathbf{J}$ on the ionic charge carriers, by superscript considered in the stream with ρ the resistivity of the stream, by subscript resulting from the electric current of density $^{str}\mathbf{J}_{str}$ induced in the stream. The $-\rho^{str}\mathbf{J}_{str}$ is the first external seawater part of the field the drifting shark detects.

$^{str}\mathbf{J}_{anl}$ denotes the induced electric current of density $\mathbf{J}$, by superscript considered in the stream, by subscript due to the drifting animal being more resistive than the stream. It is the density of the electric current forced to flow through the seawater around, instead of through the animal.

$-\rho^{str}\mathbf{J}_{anl}$ denotes the resistive force $-\rho\mathbf{J}$ on the ionic charge carriers, by superscript considered in the stream with ρ the resistivity of the stream, by subscript resulting from the electric current of density $^{str}\mathbf{J}_{anl}$ due to the resistive the animal. The $-\rho^{str}\mathbf{J}_{anl}$ is the second external seawater part of the field the drifting shark detects.

In the animal:

$^{anl}\mathbf{J}_{str}$ denotes the induced electric current of density $\mathbf{J}$, by superscript considered in the animal, by subscript originating from the ocean stream, but flowing through the drifting animal as if it were the seawater it displaces.

$-\rho^{anl}\mathbf{J}_{str}$ denotes the resistive force $-\rho\mathbf{J}$ on the charge ionic carriers, by superscript considered in the animal with ρ the resistivity of the animal, by subscript resulting from the electric current of density $^{anl}\mathbf{J}_{str}$ produced by the stream, yet flowing through the animal. The $-\rho^{anl}\mathbf{J}_{str}$ is the first internal part of the field the drifting shark detects. Taking into account that the shark is of higher resistivity than the seawater.

$^{anl}\mathbf{J}_{anl}$ denotes the induced electric current of density $\mathbf{J}$, by superscript considered in the animal, by subscript driven by the motional electric field of the electrical space charges accumulated at opposite skin surfaces of the animal, but impeded by the resistive force in the animal, and countered by the $\mathbf{v} \times \mathbf{B}$ in the animal.

$-\rho^{anl}\mathbf{J}_{anl}$ denotes the additional resistive force $-\rho\mathbf{J}$ on the ionic charge carriers, by superscript considered in the animal with ρ the resistivity of the animal, by subscript resulting from the electric current of density $^{anl}\mathbf{J}_{anl}$, in the animal, which corrects for the animal being of higher resistivity than the seawater. The $-\rho^{anl}\mathbf{J}_{anl}$ is the second, internal part of the field the drifting shark detects.

$^{anl}\mathbf{E}_{str}$ denotes the motional electric field **E**, by superscript considered in the animal, by subscript due to the electrical space charges accumulated at the boundaries of the stream.

$^{anl}\mathbf{E}_{anl}$ denotes the motional electric field **E**, by superscript considered in the animal, by subscript due to the electrical surface charges accumulated at opposite skin surfaces of the animal.

Summary Statement 3 for Shark Traveling in Still Ocean Waters

$^{anl}\mathbf{J}_{anl}$ denotes the induced electric current of density **J**, by superscript considered in the animal, by subscript driven by the $\mathbf{v} \times \mathbf{B}$ in the traveling animal, yet impeded by the resistive force in the animal, and countered by the motional electric field of the electrical space charges accumulated at opposite skin surfaces of the animal.

$-\rho^{anl}\mathbf{J}_{anl}$ denotes the additional resistive force $-\rho\mathbf{J}$ on the ionic charge carriers, by superscript considered in the animal with ρ the resistivity of the animal, by subscript resulting from the electric current of density $^{anl}\mathbf{J}_{anl}$, induced in the animal. The $-\rho^{anl}\mathbf{J}_{anl}$ is the field the traveling shark detects.

$^{anl}\mathbf{E}_{anl}$ denotes the motional electric field **E**, by superscript considered in the animal, by subscript due to the electrical surface charges accumulated at opposite skin surfaces of the animal.

Appendix 2. Mathematical Concepts of Forces and Gradients

To appreciate our elementary analysis, we have to be very precise about the physical nature of the three governing forces. The *motionally induced electric* $\mathbf{v} \times \mathbf{B}$ driving the ionic charges and the *motional electric* **E** of the resulting ionic charge accumulations are *electrical* forces in that they act on the electrical charges of the ionic charge carriers. However, the *ohmic* $-\rho\mathbf{J}$, which together with the motional electric **E** opposes the motionally induced $\mathbf{v} \times \mathbf{B}$, is not an electrical force but a *resistive drag* force of a statistical mechanics nature, acting on the charge carriers as such, not on the charges the carriers contain. Even so, $-\rho\mathbf{J}$ is treated as an electrical force in that it is defined as the resistive force per unit charge. The three forces are defined by their action on positive charges or positive charge carriers. Their action on negative charges or negative charge carriers is in the opposite direction but, in the end, has the same effect on the induced electric current of density **J**.

The motionally induced $\mathbf{v} \times \mathbf{B}$ acts as the electrical energy-imparting force. The resistive $-\rho\mathbf{J}$ acts as the electrical energy-dissipating force. The conservative motional electric field **E** of the charge accumulations acts as the electrical energy-storing reservoir, containing the energy accumulated by $\mathbf{v} \times \mathbf{B}$ minus the energy dissipated by $\rho\mathbf{J}$. The $\mathbf{v} \times \mathbf{B}$ and $-\rho\mathbf{J}$ are by definition voltage gradients, as they point energetically uphill, that is, when driving the ions in the forward direction at the cost of mechanical energy, they add electrical energy to the motional electric field $\mathbf{E}$ in the form of separating the + and – charges. The motionally induced $\mathbf{v} \times \mathbf{B}$ indeed drives the ions in their forward direction and thus tends to add energy to the motional electric field $\mathbf{E}$. The ohmic force $-\rho\mathbf{J}$ would add energy to the motional electric field $\mathbf{E}$ in the forward direction, that is, in the direction of $-\mathbf{J}$. However, being a drag force, opposing the motion of the ions, as indicated by the minus sign, it tends to subtract energy from the motional electric field $\mathbf{E}$. Let it sink in or scold me!

Conversely, the $\mathbf{E}$ field of the + and – charge accumulations is inherently the opposite of a voltage gradient, pointing energetically downhill, for it tends to bring the + and – charges closer together, thereby diminishing the electrical energy of the field. Hence, $\mathbf{E} = -\nabla\varphi$, with $\nabla\varphi$ the potential gradient of the motional electric field due to the charge separations. Unfortunately, gradients are often considered to point downhill. Indeed, in common parlor, one often speaks of a downhill gradient, even though this is contrary to the well-established mathematical convention, an unfortunate habit often found in mathematically less strict articles and textbooks.

When the situation is stationary, which means independent of time, as we initially assume, the charges may well be moving, but the charge distribution remains the same in time. Hence, we may consider the electric field $\mathbf{E}$ electrostatic, as long as we take the motion of the charges into account in the form of an electric current of density $\mathbf{J}$. However, the magnetic field that this current produces is, compared to the earth's magnetic field, so weak in our situation that we may neglect it altogether. Hence, the stationary situation is quasi-electrostatic.

In the oceanographic literature, in particular in the Physical Oceanography text by Von Arx (1962) and the theoretical treatise by Longuett-Higgins et al. (1954), which I originally started from, the physical quantities were expressed in terms of gradients, hence for the motional electric fields they used the voltage gradient $\nabla\varphi$ instead of the oppositely directed electric field strength $\mathbf{E}$. Adhering to their terminology, I consider the physically adequate stimulus for the sharks to be the pore-to-pore $\int \nabla\varphi_{\text{mem}} \cdot \mathbf{ds} = \int -\rho\mathbf{J} \cdot \mathbf{ds}$. However, when we envision the electric field $\mathbf{E}$ as the force acting on the gating charges of the electroreceptive voltage-sensitive ion channels, then, since $\mathbf{E}_{\text{mem}} = -\nabla\varphi_{\text{mem}}$, the electrical stimulus may equivalently be given by $\int \mathbf{E}_{\text{mem}} \cdot \mathbf{ds} = \int +\rho\mathbf{J} \cdot \mathbf{ds}$.

Appendix 3. Concepts of Potential Difference, Voltage, Divergence, and Curl

When we want to express the forces in terms of the work they do, that is, in terms of their line integrals from one point to another, we have to be very precise about the physical nature of these line integrals. In general, the line integrals $\int \mathbf{v} \times \mathbf{B}_{\text{v}} \cdot \mathbf{ds}$ and $\int -\rho\mathbf{J}_{\text{str}} \cdot \mathbf{ds}$ are path-dependent, so that the voltage between two points is not given by a definite number but depends on the layout of the measuring circuit. However, in the instance of an electric field **E** due to stationary charges, the line integral $\int \mathbf{E} \cdot d\mathbf{s}$ is inherently path-independent, so that the voltage between two points is a definite number, not depending on the layout of the circuit. Then, we may speak of potential differences instead of voltages between points, and we can express the electric field **E** in terms of a scalar potential field φ, where **E**, pointing down-hill, is the negative of a potential gradient, $\mathbf{E} = -\nabla\varphi$. Note: Whereas a potential gradient is a vector pointing in a definite direction, a voltage gradient is in general a second-order tensor, the size and direction of which depends on the path taken.

Furthermore, we must ascertain that the relations between the various physical quantities are consistent regarding their divergence and curl, the spatial distribution of which mathematically defines vector fields. The divergence at a point in a stationary electric field indicates the presence of a local charge accumulation, the field of which is energetically conservative, as its point-to-point line integrals are path-independent. Free of divergence or solenoidal means there are no sources or sinks. The field lines form closed loops. The curl at a point in an electric field expresses the presence of a local twist. It signifies the origin of the vorticity of the fields and renders the point-to-point line integrals path-dependent. Free of curl or irrotational means there is no vorticity or circulation. In the governing equation of the time-invariant field, $\mathbf{v} \times \mathbf{B} - \rho\mathbf{J} + \mathbf{E} = 0$, the divergence and curl must be the same on the left and the right side. Hence, since **E** has inherently no curl, $\mathbf{v} \times \mathbf{B}$ and $-\rho\mathbf{J}$ must have opposite curl. These rules are equally important as the well-known rules that not only the magnitudes but also the dimensions on the left and right side of an equation must be the same.

Appendix 4. Mathematical Consistency Concerning Divergence and Curl

This appendix addresses the mathematical consistency of the models discussed thus far. It is meant for the reader who wonders about the validity of the mathematical physics approach to motional electric orientation, in particular, whether the discourse does not violate first principles.

In the model stream, **v** and $\mathbf{B}_v$ were considered uniform and time-independent. Hence, $\mathbf{v} \times \mathbf{B}_v$ was free of curl and free of divergence under the assumed conditions. The electrostatic $\mathbf{E}_{str}$ is inherently free of curl, so that, according to the governing equation $\mathbf{v} \times \mathbf{B}_v - \rho\mathbf{J}_{str} + \mathbf{E}_{str} = 0$, the $-\rho\mathbf{J}_{str}$ and, at homogeneous resistivity, $\mathbf{J}_{str}$ were free of curl as well. Furthermore, in the steady state, $\mathbf{J}_{str}$ is inherently free of divergence, so that, at homogeneous resistivity, $-\rho\mathbf{J}_{str}$ and $\mathbf{E}_{str}$ were free of divergence as well. However, at the boundaries of the stream, where **v** was discontinuous, or rather changed rapidly, neither $\mathbf{v} \times \mathbf{B}_v$ nor $\mathbf{E}_{str}$ were free of divergence, actually their divergences canceled according to the governing equation. Furthermore, neither $\mathbf{v} \times \mathbf{B}_v$ nor $-\rho\mathbf{J}$ were free of curl, actually their curls also canceled according to the governing equation. It was the divergences and curls that led to the flow of motionally induced electric current around the circuit across the stream and back through deeper water and the ocean floor with respect to which the stream moves. Or, stated otherwise, the fields had their origin at the boundaries of the stream, even though the $\mathbf{v} \times \mathbf{B}_v$ of the wind-driven stream supplied the energy, which was stored in the **E** field of the charge accumulations at the boundaries of the stream and dissipated by the $-\rho\mathbf{J}$ along the current paths through the stream and through deeper water and the ocean floor.

As for the shark interacting with the earth's magnetic field when moving relative to the water, in our model, **v** the velocity of the shark and $\mathbf{B}_h$ the horizontal component of the earth's magnetic field were considered uniform and time-independent. Hence, $\mathbf{v} \times \mathbf{B}_h$ was free of curl and free of divergence. Since by definition $\mathbf{E}_{anl}$ due to the charges on opposite skin surfaces, measured in the animal, and $\mathbf{E}_{rta}$ due to the same charges, measured along the return paths through the seawater medium, are free of curl, according to the governing equations $\mathbf{v} \times \mathbf{B}_h - \rho\mathbf{J}_{anl} + \mathbf{E}_{anl} = 0$ and $-\rho\mathbf{J}_{rta} + \mathbf{E}_{rta} = 0$, also $-\rho\mathbf{J}_{anl}$ in the animal and $\rho\mathbf{J}_{rta}$ in the surrounding water were free of curl. Hence, at homogenous resistivity, also, $\mathbf{J}_{anl}$ and $\mathbf{J}_{rta}$ were free of curl. Similarly, $-\rho\mathbf{J}_{anl}$ and $-\rho\mathbf{J}_{rta}$ were in the stationary state free of divergence. Hence, inside the animal also $\mathbf{E}_{anl}$ and $\mathbf{E}_{rta}$ were free of divergence as well. However, since in reality the resistivity of the skin and body tissues is not homogeneous, $\mathbf{J}_{anl}$ in general is not free of curl and $\mathbf{E}_{anl}$ is not free of divergence, which however does not affect our reasoning, since $-\rho\mathbf{J}_{anl}$, the line integral of which the animal detects, remains free of curl. At the surface of the shark, where the velocity **v** was discontinuous, neither $\mathbf{v} \times \mathbf{B}_v$ nor $\mathbf{E}_{anl}$ were free of divergence, and neither $\mathbf{v} \times \mathbf{B}_v$ nor $-\rho\mathbf{J}$ were free of curl. Hence, mathematically, it was their curl and divergence that led to the circulation of the current around the closed circuit of the shark and the surrounding seawater.

Appendix 5. Detailed Analysis of a Technical Example

Although everybody knows how to measure the voltage of a lead-acid battery, do we really know what we are doing? In daily life, one may not need to, but in studying the electric sense of sharks and rays, one had better make a sincere effort thoroughly to understand this intricate process. The detailed analysis of this technical example may be very helpful, in the hope that it does not offend the reader. It certainly took the author a while to get it straight, if he really did.

Consider an unloaded lead-acid battery. Electrochemical forces $\mathbf{F}_{chem}$ drive the positive ions from the negative to the positive plate and the negative ions from the positive to the negative plate until the resulting charge separation produces the counteracting electrostatic forces $\mathbf{E}_{elec}$ needed to balance the electrochemical forces $\mathbf{F}_{chem}$, that is, until $\mathbf{E}_{elec} + \mathbf{F}_{chem} = 0$ in terms of forces or $\nabla\varphi_{elec} = \nabla\varphi_{chem}$ in terms of potential gradients. In the stationary state, the work to be done *by* the electrochemical forces to move positive charges from the negative to the positive plate, that is, the line integral of the electrochemical forces from the negative to the positive plate,

$\int \mathbf{F}_{\text{chem}} \cdot \mathbf{ds}$ or $\int \nabla \varphi_{\text{chem}} \cdot \mathbf{ds}$, gives the electrical potential or emf of the battery. By the same token, the work done *on* the electrical forces, when the positive charges are moved from the negative to the positive plate, that is, the line integral of the electrical forces from the negative to the positive plate, $\int -\mathbf{E}_{\text{elec}} \cdot \mathbf{ds}$ or $\int \nabla \varphi_{\text{elec}} \cdot \mathbf{ds}$, naturally also gives the electrical potential or emf of the battery. Hence the potential of the battery is $\int \nabla \varphi_{\text{chem}} \cdot \mathbf{ds} = \int \nabla \varphi_{\text{elec}} \cdot \mathbf{ds}$, both of which are the energetically uphill line integrals of potential gradients.

Next connect an ideal voltmeter to the terminals of the battery. Whereas in the battery the electrochemical and electrostatic forces balance, in the voltmeter there are no electrochemical forces. Hence, in the voltmeter, some other forces must match the electrochemical forces of the battery. Due to the electrochemically produced potential of the battery, initially electronic charges flow through the test leads into the voltmeter. If we consider the sensing element of the voltmeter to act as a pure dielectric, then positive and negative charges will build up on either side of the sensing element and along the wall of the two test leads until they null the electric fields in the test leads. In the sensing element, these charge accumulations cause polarization forces, $\mathbf{P}/\varepsilon\chi$, which are kept in balance by the dielectric's internal counteracting forces, $-\mathbf{P}/\varepsilon\chi$, the line integral of which, $\int -\mathbf{P}/\varepsilon\chi \cdot \mathbf{ds}$, is the equivalent of the line integral of the electrochemical forces in the battery, $\int \mathbf{F}_{\text{chem}} \cdot \mathbf{ds}$. Hence, ultimately it is the resulting $\int -\mathbf{P}/\varepsilon\chi \cdot \mathbf{ds}$ that the voltmeter detects.

When the battery gets loaded and its positive ions travel at constant velocity to the positive plate and its negative ions to the negative plate, the forces balancing the electrochemical forces are no longer only the electrostatic forces due to the charge accumulations at the plates, but also the resistive forces, $-\rho\mathbf{J}$ in the battery. Hence, for the loaded battery $\int \nabla \varphi_{\text{elec}} \cdot \mathbf{ds} = \int (\nabla\varphi_{\text{chem}} - \rho\mathbf{J}) \cdot \mathbf{ds}$, as a result of the internal resistance. The example gets close to the shark's electroreceptor system, except where the electrical stimuli do not arise from electrochemical, but from motional electric forces, the topic of discussion of the present article.

Appendix 6. Axiom: Electric Field Inside and in the Vicinity of a Resistive Sphere

Envision a vast open-ocean volume of seawater invaded by a uniform ambient electric current field. Introduce a uniformly resistive spherical body into the ambient electric current field. If the resistivity of the body is higher than that of the seawater, + and – electrical charges will accumulate on the fore and aft sides of the sphere, causing part of the electric current to be diverted around the sphere. The more resistive the body is, the larger the charge accumulations are, the more of the electric current diverts around the sphere, and the less of the electric current flows through the sphere.

Mathematical analysis shows that the electric current field *inside* a uniformly resistive sphere introduced into a uniform ambient electric current field is uniform as well, although less in strength when the sphere is more resistive than the seawater

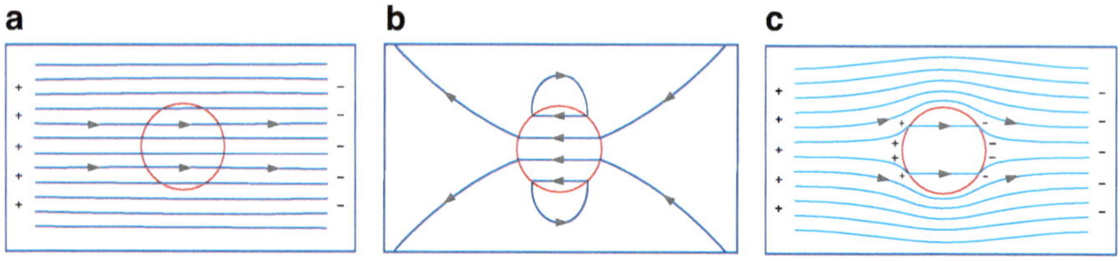

Fig. 2.10 (**a**) Seawater sphere, (**b**) adjusting current field, (**c**) resistive sphere

it displaces. The weaker uniform electric current field in the sphere may be considered to consist of the uniform electric current field in the undisturbed seawater *plus* an *oppositely* directed uniform electric current field of lesser strength, adjusting the internal uniform electric current field in agreement with the resistivity of the sphere.

The electric current field *outside* the sphere may be considered to consist of the original uniform electric current field in the undisturbed seawater *plus* an electric dipole current field, the virtual source of which is located at the center of the sphere, impelling electric current to flow through the seawater outside the sphere from the + charges on the fore side to the – charges on the aft side of the sphere. The oppositely directed uniform electric current field inside the sphere adjusting the reduced total electric current in the sphere and the matching dipolar electric current field outside the sphere form a continuous closed loop current through the sphere and back through the seawater medium (Fig. 2.10).

Physically, the vital differences between the seawater sphere (left panel) and the more resistive sphere (right panel) are the + and – charge accumulations on the fore and aft surfaces of the sphere and the lesser electrical current density inside the sphere. We will speak of the sphere's adjusting electric current field (middle panel), that is the electric current field that describes the change in the situation from before to after introduction of the resistive sphere into the undisturbed uniform electric current field. Note that the sphere's electric current field is not created by the sphere but represents the change in the ambient uniform electric current field due to the introduction of the resistive sphere.

To prepare for the application to a shark drifting in an ocean stream, we will use the subscript *str* for the undisturbed ambient field, even when the water is not flowing, *anl* for the added current dipole field inside and outside the spherical body, while the superscript indicates where the field is measured, *anl* inside the sphere and *str* in the water outside the sphere.

In the Undisturbed Ambient Electric Current Field The uniform electric current density ${}^{str}\mathbf{J}_{str} \rightarrow$ of the stream is in the stream associated with a uniform voltage gradient ${}^{str}\nabla\varphi_{str} = -\rho{}^{str}\mathbf{J}_{str} \leftarrow$, with ρ the resistivity of the seawater.

In the Resistive Sphere the uniform electric current density of the stream ${}^{anl}\mathbf{J}_{str} \rightarrow$ is in the sphere associated with a uniform voltage gradient ${}^{anl}\nabla\varphi_{str} = -\rho{}^{anl}\mathbf{J}_{str} \leftarrow$, with ρ the resistivity of the sphere. The adjusting uniform current of the sphere ${}^{anl}\mathbf{J}_{anl} \leftarrow$ is in sphere associated with a uniform voltage gradient ${}^{anl}\nabla\varphi_{anl} = -\rho{}^{anl}\mathbf{J}_{anl} \rightarrow$, with ρ the resistivity of the sphere. Adding the two yields in the sphere ${}^{anl}\mathbf{J}_{str} \rightarrow +{}^{anl}\mathbf{J}_{anl} \leftarrow$ and ${}^{anl}\nabla\varphi_{str} \leftarrow +{}^{anl}\nabla\varphi_{anl} \rightarrow$ and $-\rho^{anl}\mathbf{J}_{str} \leftarrow + - \rho^{anl}\mathbf{J}_{anl} \rightarrow$, with ρ the resistivity of the sphere.

In the Vicinity of the Resistive Sphere The adjusting dipole sphere return current of density ${}^{str}\mathbf{J}_{anl} \leftarrow \rightarrow \leftarrow$ is associated with a dipolar voltage gradient ${}^{str}\nabla\varphi_{anl} = -\rho^{str}\mathbf{J}_{anl} \rightarrow \leftarrow \rightarrow$, with ρ the resistivity of the seawater. Adding this to the field existing in the undisturbed seawater in the vicinity of the sphere yields ${}^{str}\mathbf{J}_{str} \rightarrow +{}^{str}\mathbf{J}_{anl} \leftarrow \rightarrow \leftarrow$ and ${}^{str}\nabla\varphi_{str} \leftarrow +{}^{str}\nabla\varphi_{anl} \rightarrow \leftarrow \rightarrow$, and $-\rho^{str}\mathbf{J}_{str} \leftarrow + \rho^{str}\mathbf{J}_{anl} \rightarrow \leftarrow$.

Note: of the sets of three arrows, the middle arrow applies to the field near the vertical symmetry plane of the sphere, where the field is stronger than undisturbed, whereas the first and third arrows apply to the field to the left and right of the sphere, where the field is weaker than undisturbed.

The voltage difference between the fore and aft apices of the resistive sphere, given by the line integral $\int({}^{str}\nabla\varphi_{str}+{}^{str}\nabla\varphi_{anl}) \cdot \mathbf{ds}$ along an external path through the seawater surroundings of the sphere, is mathematically determined to be 1–1.5 times $\int{}^{str}\nabla\varphi_{str} \cdot \mathbf{ds} = \int -\rho^{str}\mathbf{J}_{str} \cdot \mathbf{ds}$ of a similar path through a sphere, depending on the resistivity of the sphere ranging from that of the seawater to infinitely high. The voltage between the fore and aft apices of the resistive sphere is also given by the line integral along an internal path through the sphere $\int({}^{anl}\nabla\varphi_{str}+{}^{anl}\nabla\varphi_{anl}) \cdot \mathbf{ds} = \int(-\rho^{anl}\mathbf{J}_{str} + \rho^{anl}\mathbf{J}_{anl}) \cdot \mathbf{ds}$. Hence, since the voltage between the apices is the same for the external and the internal line integrals, the outside $\int(-\rho^{str}\mathbf{J}_{str} - \rho^{str}\mathbf{J}_{anl}) \cdot \mathbf{ds}$ equals the inside $\int(-\rho^{anl}\mathbf{J}_{str} + \rho^{anl}\mathbf{J}_{anl}) \cdot \mathbf{ds}$. Hence, the inside $\int(-\rho^{anl}\mathbf{J}_{str} + \rho^{anl}\mathbf{J}_{anl}) \cdot \mathbf{ds}$ also equals 1–1.5 times $\int{}^{str}\nabla\varphi_{str} \cdot \mathbf{ds} = \int -\rho^{str}\mathbf{J}_{str} \cdot \mathbf{ds}$ of a similar path through the undisturbed seawater.

Similar relations hold for ellipsoid bodies. For arbitrarily shaped bodies though, the field inside the body is no longer uniform, even when the ambient field is. However, to keep the analysis as simple as possible, but not simpler, we will restrict ourselves to spherical ellipsoids. The treatise is a matter of physical principles, not of the gory details.

The axiom applied to how a shark redistributes the electric current field in ocean waters did not specify the source of the oceans' uniform electric current field. It was merely indicated by + and – signs on the far left and right sides of the field. Yet, this is exactly the situation a stationary shark perceives in its own rest frame of reference. However, in the present article, we discuss the situation as seen by the earthbound observer, who considers the $\mathbf{v} \times \mathbf{B}$ of the flowing water the driving force. Rest assured, though, the effect of the shark on the electric current is the same in both cases, as in the shark frame the $\mathbf{v} \times \mathbf{B}$ drops out but reenters the equation in the guise of an additional imposed electric field $\mathbf{E}$ by remote charges as inferred by Einstein (1905).

When in an open-ocean stream the uniform electric current field is driven by the $\mathbf{v} \times \mathbf{B}$ rather than the + and – charges at the lateral boundaries of the stream, but the sphere remains stationary, the governing equation in the stream is $\mathbf{v} \times \mathbf{B}_v - \rho\mathbf{J}_{str} = \nabla\varphi_{str}$ rather than $-\rho\mathbf{J}_{str} = \nabla\varphi_{str}$. We may analyze three situations: (1) the stream is virtually shorted by the return current, that is, $\nabla\varphi_{str} = 0$, (2) the stream is not loaded at all, that is, $-\rho\mathbf{J}_{str} = 0$, and (3) the stream is moderately loaded, $-\rho\mathbf{J}_{str} = \mathbf{v} \times \mathbf{B}_v - \nabla\varphi_{str}$.

As for (1), whether the uniform electric current in the stream is driven by a uniform electric field produced by imposed + and – charges at the lateral boundaries of the stream or by a uniform $\mathbf{v} \times \mathbf{B}_v$ field of similar strength makes no difference as to how the current density $\mathbf{J}_{str}$ is affected by the resistive sphere. As for (2), the non-existing current density $\mathbf{J}_{str}$ is not affected by the resistive sphere. As for (3), the electric current of density $\mathbf{J}_{str}$ driven by $\mathbf{v} \times \mathbf{B}_v$ is now counteracted by the opposing $\nabla\varphi_{str}$ of the – and + charges at the lateral boundaries of the stream it creates. The results of the first and third cases are that the $-\rho\mathbf{J}_{anl} = 1 - 1.5$ times $-\rho\mathbf{J}_{str.}$

In review, the electric current density ${}^{str}\mathbf{J}_{str}$ in the undisturbed seawater would also be the electric current density in the shark if the animal would have the same resistivity as the seawater. However, since the average 2 kohm cm resistivity of the animal is higher than the 20 ohm cm resistivity of the seawater, the current density in the animal equals, according to the axiom, the current density ${}^{str}\mathbf{J}_{str}$ in the seawater *plus* an oppositely directed adjusting current density ${}^{str}\mathbf{J}_{anl}$ due to the presence of the resistive animal. The return current of density ${}^{str}\mathbf{J}_{anl}$ in the seawater outside the sphere, which is continuous with the added adjusting current of density ${}^{anl}\mathbf{J}_{anl}$ in the sphere, drives the electric current around the animal through the seawater environment, adding to the voltage across the sphere.

References

Carey FG, Scharold JV with appendix by Kalmijn AJ (1990) Movements of blue sharks (*Prionace glauca*) in depth and course. Mar Biol 1006:329–342

Einstein A (1905) Zur Elektrodynamik bewegter Körper. Annalen der Physik 17:891–921. See also: On the electrodynamics of moving bodies. In: The principle of relativity. Translated by W. Perrett and G. B. Jeffery, 1952, Dover edition, pp 35–65

Föppl A (1904, 1907) Einführung in die Maxwellsche Theorie der Elektrizität. Druck und Verlag von B. G. Teubner, Leipzig

Longuett-Higgins MS, Stern ME, Stommel H (1954) The electric field induced by ocean currents and waves, with applications to the method of towed electrodes. Papers in Physical Oceanography and Meteorology, Massachusetts Institute of Technology and Woods Hole Oceanographic Institution. Contribution No 690

Purcell EM (1985) Electricity and magnetism, 2nd edn. McGraw Hill, New York

Purcell EM, Morin DJ (2013) Electricity and magnetism, 3rd edn. Cambridge University Press, New York

von Arx WS (1962) An introduction to physical oceanography. The electromagnetic method. Adison-Wesley, pp 260–281

Waltman B (1966) Electrical properties and fine structure of the ampullary canals of Lorenzini. Acta physiol scand 264:1–60

Chapter 3
Sharks Exploring Ambient Electric and Earth's Magnetic Fields

3.1 Exploratory Movements Integral Part of Sensory System

Introduction
Imagine a shark traveling in an open-ocean stream in the presence of the earth's magnetic field, as a result being exposed not only to the motional electric field of the flowing water but also to the electric field it induces by moving with respect to the water. Might the animal electrically sense the strength and direction of the stream as well as its own magnetic compass heading? We will address the issue under idealized, readily understood conditions, yet remove the constraints after affirming the physical and behavioral feasibility of the idealized situations. Even so, we will not endeavor to *establish* how the animals orient electrically and magnetically, but we will *assume* what we motion-electrically can achieve imagining to be sharks, real sharks can achieve more sensibly.

In the author's initial accounts of nearly half a century ago, the motion-electric fields sharks *receive* and the exploratory movements they may perform to *detect* the electric and magnetic fields were outlined only in principle. The reader must realize that, in these pioneering years, he wondered but did not know whether sharks might orient to the motional electric fields of ocean currents and did not know whether sharks would be able to use their electric sense to monitor their magnetic compass headings. At that early stage, it was purely a matter of inference based upon logical reasoning and physical contemplation, starting from the astounding electrical sensitivity of sharks and the kindred rays he had just established behaviorally. Only after assuring the biological and physical feasibility of his motional electric orientation theory, he felt ready to seek behavioral evidence of the electric and magnetic orientation of sharks and rays, as he successfully did in subsequent years.

From the very start, the author has been fully aware that the electroreceptors of sharks and rays, the ampullae of Lorenzini, require the animals to perform the proper exploratory movements not only to bring the oceans' DC electric and magnetic fields into the low-frequency AC range of their electroreceptors, the ampullae

A. J. Kalmijn, *Theory of Electric and Magnetic Orientation in Sharks and Rays Revisited*, https://doi.org/10.1007/978-3-031-58445-9_3

of Lorenzini, but also to interpret the motional electric fields they receive in terms of electric and magnetic orientational cues. Even zeroing in on DC dipole fields simulating the bioelectric fields of prey requires relative motion between the shark and its target. Therefore, the objective of the present section is to conceive of the exploratory movements sharks and rays may make to acquire the sensory information needed to orient in the worldwide ambient electric and magnetic fields. This certainly is a realistic effort, since the motion-electric fields ocean currents and the sharks themselves induce during their normal behavior are constantly exciting the ampullary sense organs well within their dynamic and frequency range, as I learned from both behavioral and afferent-nerve responses in free-swimming sharks.

A touchy issue

To render the present section independent of the foregoing scholarly discourse, we will base the discussion on the pared-down version of the magnetic orientation theory, given that its limited validity and the approximations made are well understood. That is, we will, with great trepidation, consider the shark electrically to be all but shorted by the seawater medium and, accordingly, not take into account the small motionally induced charge accumulations on opposite skin surfaces of the animal, despite my dire warning that, strictly speaking, the pared-down version pertains only to the limit of a mathematical situation that physically can-never be reached. Nevertheless, presently we will use it as an expedient but treacherous aid for the convenience of the reader, justifiable as it does not affect the validity of the conclusions we will arrive at. Yes, I am ambivalent, first I make my own theory overly simple for the layman but then scold myself for doing so, afraid of competing with my comprehensive theory. To make it easier for the reader, we will also analyze the shark's movements as we observe them relative to our familiar earth frame of reference, yet all the time keeping in mind the electrical signals the shark receives in its own frame of reference on the basis of which it does the orienting.

Physically speaking, the motionally induced electric $\mathbf{v} \times \mathbf{B}$ field is the driving force. Nevertheless, sharks that motion-electrically sense their magnetic compass headings do not detect the $\mathbf{v} \times \mathbf{B}$ but the resulting resistive $-\rho\mathbf{J}$, as explained in the former section. The difference between the two, the driving $\mathbf{v} \times \mathbf{B}$ and the resistive $-\rho\mathbf{J}$, is the motional electric field $\mathbf{E}$ of the electrical charges accumulating on opposite skin surfaces of the animal. However, since marine sharks are electrically nearly shorted by the seawater medium, the charge accumulations on the skin and their motional electric field $\mathbf{E}$ are only small—but never zero—so that $-\rho\mathbf{J}$ approaches $-\mathbf{v} \times \mathbf{B}$,—but never becomes $-\mathbf{v} \times \mathbf{B}$, as the $-\mathbf{v} \times \mathbf{B}$ and $-\rho\mathbf{J}$ are physically very different quantities. Hence, under these limiting conditions, we may use $\mathbf{v} \times \mathbf{B}$ as a *proxy* for $-\rho\mathbf{J}$ without deviating significantly from the real situation in the instance of marine sharks. The pared-down version has the advantage that the driving $\mathbf{v}$ and $\mathbf{B}$ are more familiar and more instructive physical entities than the resulting $-\rho\mathbf{J}$ the animal detects. However, when a shark enters fresh waters, the charge accumulations and, hence, $\mathbf{E}$ are no longer small and cannot be neglected. In general, with increasing resistivity of the medium relative to which the animal moves, from seawater to thin air, the proper stimulus $-\rho\mathbf{J}$ may vary from the full $-\mathbf{v} \times \mathbf{B}$ to nil. Note, where I said $\mathbf{v} \times \mathbf{B}$, $\mathbf{E}$, or $-\rho\mathbf{J}$, I tacitly meant the pore-to-pore line integrals of them.

Remember and foresee

The details of the following discourse may be strenuous, but remember from the former section: (1) For sharks, the electric fields they receive from their drift with ocean streams and from their own motional activity are very similar, both are essentially uniform. (2) The electric signals sharks receive from the ambient electric and the earth's magnetic fields are readily sensed side by side, as they are different functions of the animals' exploratory movements. (3) The ambient electric fields are *imposed* upon the animals, though the electrical signals they receive from them are *under the control of* their own motional activity, whereas the electrical signals sharks receive from interaction with the earth's magnetic field are the *result of* their own motional activity. (4) Ampullary excitations fade away with a time constant of about 4 s, long enough to render the system virtually DC-sensitive for brief signals, but short enough to reset the excitation within seconds. (5) By assumption, sharks detect the electric fields three-dimensionally, rendering them capable of focusing on the components of the fields they receive along the mutually orthogonal aft-fore, ventro-dorsal, and left-right body axes.

We will start from the individual exploratory moves sharks—I *assume*—make to orient in the oceans' ambient electric and magnetic fields. Subsequently, we will string the individual moves together, first pairwise and then in longer series, to study the signals the animals receive when haphazardly moving about and when traveling steadily. Note, however, that for them it is not only of importance constantly to monitor the ambient electric and magnetic fields for open-ocean orientation but also strictly necessary in order to take the ambient electric and magnetic fields into account when zeroing in on the bioelectric fields of their prey, amazing achievements on the part of the predators. The present section endeavors to envision the biologically realistic exploratory moves that enable sharks concurrently to sense ambient electric and magnetic fields. I will strive strictly to adhere to the governing laws of physics and keep in mind the biological capabilities of the animals. In short, I will make an effort to prove the feasibility of how the animals might operate. Whether the animals really operate the way I envision is a question left for the animals and future researchers.

3.2 Orienting in DC Ambient Electric and Magnetic Fields with AC Receptors

For readers who have been wondering how sharks orient in DC electric and magnetic fields by use of their low-frequency AC receptors, let us begin with two thought experiments to introduce the issues in their most elementary forms in preparation of more detailed analyses. A reader who is unsure of the following two paragraphs shall continue reading the discourse and refer back now and then until both paragraphs make perfect sense. The two introductory thought processes may be tedious but are perfectly logical (I hope). In the subsequent formal discourse, we will look into the various situations in greater detail so as gradually to get more familiar with the exploratory moves enabling sharks and rays to orient electrically as well as magnetically. However, do not wonder how the animals physically and mathematically analyze the various situations. For them, it is purely a matter of associating the proper behavioral responses with the particular self-initiated stimulus patterns, however complicated we as scientists may describe them. Even when walking a tightrope, we operate in a similar fashion, without trying physically and mathematically to understand how we do it. Yet, we can only succeed if it is mathematically and physically feasible.

Sensing direction and strength of horizontal component of local ambient electric field
Let a shark, after coasting rectilinearly in a level plane long enough to reset its electroreceptors, turn at constant forward speed right or left through an arbitrary angle in a time short compared to the adaptation time of the sense organs. The direction along which the horizontally oriented ampullary sense organs receive the smallest signals then indicates the direction of the horizontal component of the ambient electric field, whereas the strength of the signals received along the direction normal to the ambient field give the animal a measure of the strength of the ambient electric field. Interaction of the turns with the horizontal component of the earth's magnetic field are dorso-ventrally directed, hence they do not interfere with the detection of the horizontal ambient electric field. More intricately, interaction of the horizontal turning motions with the vertical component of the earth's magnetic field produces at opposite sides of the animal oppositely directed motional electric fields that differentially cancel one another. Hence, neither of them interferes with the horizontal ambient-electric field signals received. Turns of a mere 3 degrees would suffice to orient in an ambient electric field of 100 nV/cm.

Sensing direction and strength of horizontal component of local earth's magnetic field
Let a shark, after drifting a while to reset its electroreceptors, execute a horizontal acceleration spurt reaching an arbitrary reproducible speed, followed by an instant of a rectilinear constant-velocity coasting before decelerating again, all in a time short compared to the adaptation time of its electroreceptors. By executing this exploratory behavior, let the shark empirically determine the direction in which, during the instant of coasting, it does NOT receive a ventro-dorsal signal from interaction with the horizontal component of the earth's magnetic field, that is when moving due north or due south. Let the animal then turn 90 degrees and, after again drifting a while to reset the sense organs again, once more execute the acceleration spurt. The strength of ventro-dorsal signal the shark now receives during the instant of coasting gives it a measure of the strength of the horizontal component of the magnetic field, while the polarity of the signal informs the animal whether it faces due east or due west, in which directions even spurts of a mere 2 cm/s would suffice in 20 μT fields. In addition to sensing the cardinal directions E, W, N, and S by this procedure, the consecutive rectilinear moves give the animal left-right signals providing a measure of the vertical component of the earth's magnetic field. The ratio

of the horizontal and vertical components gives the animal the local inclination or dip of the earth's magnetic field. During the instance of coasting, neither the ventro-dorsal nor the left-right signals are affected by the presence of a uniform ambient electric field.

3.3 Magnetic Compass Orientation, Rectilinear Acceleration Procedure

To start the formal discourse, we will elaborate on the concise introductory statements. As above, imagine a shark to accelerate with a few tailbeats in a horizontal direction up to a reproducible speed and then to coast *rectilinearly* at constant velocity for an instance short compared to the adaptation time of the sense organs, cf. Fig. 1.1a–c. By interacting with the horizontal component of the earth's magnetic field, it then receives during the coasting a vertical electric signal along its ventro-dorsal body axis and by interaction with the vertical component of the earth's magnetic field a horizontal electric signal along its left-right body axis. The signals it thus receives are proportional to the speed of coasting, to the strength and polarity of the magnetic field, and most importantly, for the ventro-dorsal signals, to the sine of the angle θ that its velocity makes with the horizontal component of the magnetic field, whereas the motional electric signals it receives along the left-right axis due to interaction with the vertical component of the magnetic field are independent of the animal's compass headings. Let the strengths of the ventro-dorsal signals be measured in proportion to the strongest signals received during the accelerations, that is, when the animal moves due east or due west.

On further contemplation, it becomes evident that, by executing the accelerations in two different directions, say 30 degrees apart, the shark receives (1) from the swimming direction-dependent ventro-dorsal signals unambiguous information about the direction of the horizontal component of the earth's magnetic field, hence of its magnetic compass heading, (2) from the swimming direction-independent left-right signals a measure of its to be reproduced speed, (3) from the relative strength of the ventro-dorsal and left-right components the inclination of the earth's magnetic field, hence its magnetic latitude. In short, the shark may sense—*what we call*—the strength and direction of the earth's magnetic field. It only requires the animal to execute in strength reproducible accelerations in two different directions and to be able to compare the results of consecutive moves. Note that the second move is necessary to remove all ambiguity. To maintain a given compass heading, the animal may just keep the signals received from successive accelerations constant. Be aware, the above scenarios are described as a sequence of discrete behavioral phases for the sake of clarity, but the logic pertains to the more natural, smooth behavioral moves just as well and may be recognized as elements of the seemingly haphazard maneuvers the animals perform in daily life. For detailed accounts, see the legends of Fig. 1.1a–c (from Kalmijn 1997) and Fig. 1.2 (from Kalmijn 1981). Yet, I must warn the reader that it is not easy conceptually to follow the logical

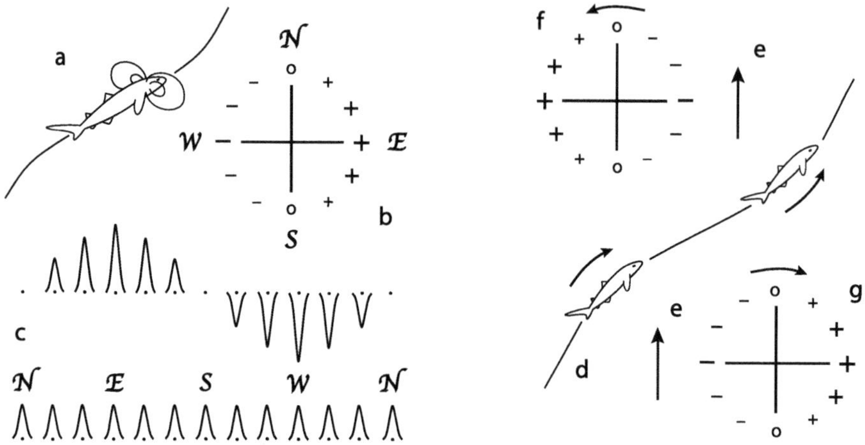

Fig. 1.1 Left: When a shark, traveling in a horizontal plane, *rectilinearly* accelerates and decelerates again in different directions, it *receives*, by interaction with the horizontal component of the earth's *magnetic* field, transient, upward-directed, ventro-dorsal (+) or downward-directed, dorso-ventral (−) electric signals, from which it may derive its magnetic compass heading. (**a**) Path of shark traveling on a rectilinear stretch, (**b**) strength and direction of ventro-dorsal signals received as a function of compass heading, (**c**) strength and direction of ventro-dorsal signals (top) and strength and direction of left-right signals as well as of inner-ear linear acceleration stimuli (bottom). The relative strengths of the ventro-dorsal and left-right signals along the east-west axis also give the animal its magnetic latitude on the earth's globe. Notice that each pair of successive responses to 30-degree apart rectilinear accelerations is unique, taking the direction of turning into account, while ambient-electric fields do not affect the animal during the rectilinear moves. Right: When a shark makes a left or right turn, *at constant forward speed*, in a uniform, horizontal, ambient *electric* field, it *detects* an, in strength increasing, electric signal rotating at half the rate the animal turns (see Fig. 1.4), yet in the opposite direction. (**d**): Right turn followed by left turn, (**e**): direction of imposed ambient-electric field, (**f**): potential distribution in terms + and − potentials received by shark directly after start of left turn, (**g**): transient potential distribution in terms + and − *potentials* received by shark directly after start of right turn. Note that at the start of the turn the signal *detected* is perpendicular to the field *imposed*, to the right upon a left turn and to the left upon a right turn. Hence, the parting line between positive and negative potential regions is *initially* parallel to the ambient-electric field, in the direction of the field when the positive region on turning right is on the right side, and *vice versa*. Note that, during the constant forward speed, the vertical signal due to interaction with the horizontal component of the earth's magnetic field does not interfere with the horizontal ambient-electric field detection, and that the turning moves by interaction with the vertical components of the earth's magnetic field induce on opposite sides of the body oppositely directed electric signals that cancel one another

reasoning, as I myself have painfully experienced upon proofreading after a long while.

But how might the animal keep the electrical signals received from moving in the earth's magnetic field apart from the signals received from the ambient electric field? *When the animal accelerates rectilinearly in different compass directions, it does not receive a signal from the ambient electric field, neither from the horizontal nor from the vertical component, only from interaction with the earth's magnetic field, with the horizontal as well as the vertical component. Hence, sensing its*

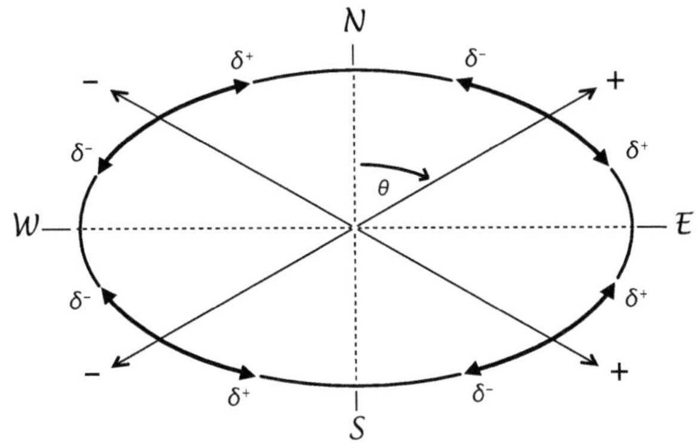

Fig. 1.2 Shark exploring compass headings *by accelerating and subsequently decelerating rectilinearly* in different directions. The strength of the ventro-dorsal electric signal is proportional to the sine of the angle θ, which is the same in four directions on the compass rose, e.g., NE, SE, SW, NW. By taking the sign of the signal into account, that is, whether the signal is ventro-dorsal (+) or dorso-ventral (−), this reduces to only two directions, NE or SE and SW or NW, respectively. The selection between these pairs of directions may be made by considering the change δ in the strength of the signal after turning into an adjacent direction. Whether the strength of the signal increases or decreases after turning into an adjacent direction depends on the + or − sign of the derivative of the sine, that is, on the sign of the cosine of the angle θ. Thus, when the sign of the cosine is positive the signal becomes stronger, from (−) toward (+), when the turn is to the right, and weaker, from (+) toward (−), when the turn is to the left, as in the NE and NW directions. *Vice versa,* when the sign of the cosine is negative the signal becomes weaker when the turn is to the right and stronger when the turn is to the left, as in the SE and SW directions. Hence, taking the increase δ^+ or decrease δ^- of the signal after turning into account unambiguously reduces the selection to just one, unique magnetic compass heading, depending on the strength of the signal, the sign of the signal, and the sign of δ. Let E_{rel} be the strength of the signal divided by the strength when the animal accelerates and decelerates due east, then, expressed logically, the angle θ is given by $(\theta = \sin^{-1}E_{rel} \text{AND} \, \delta > 0)$ OR$(\theta = \sin^{-1}E_{rel} \text{AND} \, \delta < 0)$. When successive moves straddle the E or W directions, the amplitudes of the signals are strongest, and the sign of the cosine changes. When successive moves straddle the N or S directions, the amplitudes of the signals are weakest, whereas the sign of the cosine remains the same. To include larger turns, the signal may be calculated to any desired precision by developing the sine of $\theta + \delta$ in a Taylor series: $\sin(\theta + \delta) = \sin\theta + 1/2\,\delta \cos\theta + 1/6\,\delta^2 \sin\theta + ..$ with θ the angle of the first exploratory move, and $\theta + \delta$ the angle of the subsequent exploratory move. Note that with the signals the shark receives, we mean the signals in response to its exploratory motions, its linear accelerations or decelerations

magnetic compass heading in this manner is not affected by the ambient electric field at all. The procedure of rectilinearly accelerating in different directions is, in our view, conceptually the clearest method for a shark to orient magnetically in the presence of the ambient-electric field. Yet, it does not necessarily mean that it is the preferred method for the animal. There are alternative unambiguous methods, as we will see below. Even though by rectilinearly accelerating sharks are not stimulated by the ambient-electric field, this should not be construed as ambient electric fields to be of less importance. To the contrary, the author has proved that they do use the ambient-electric field to orient in familiar waters. Actually, both in conditioning experiments in land-based seawater habitats and in tests in the wild, the animals followed the experimentally applied changes in the direction of the ambient *electric* field, even though the ambient *magnetic* field was not altered, indicating that during their travels the ambient electric fields took priority over the earth's magnetic field. This all follows directly from the physical principles outlined in my Kalmijn (1974) and later articles.

How strong are the transient electrical signals a rectilinearly accelerating and decelerating shark receives by interaction with the horizontal component of the earth's magnetic field? When, after adaptation of the sense organs, the animal accelerates to a speed of 1 m/s, within a time short compared to the adaptation time constant, the ventro-dorsal signal it receives varies as a sine function from 200 nV/cm when accelerating along the E–W axis to zero along the N–S axis at a temperate-zone, horizontal magnetic field of 20 μT. Upon making 30-degree turns, the change in the signals ranges from a maximum of about 100 nV/cm in northern and southern directions to a minimum of 7 nV/cm in eastern or western directions. Hence, during the animal's normal behavior, the method of rectilinear accelerations and decelerations is effective over its full range of compass headings. Actually, sharks may speed up in bursts of 10 m/s, receiving signals tenfold stronger, from a maximum of 1000 to a minimum of 70 nV/cm.

3.4 Ambient Electric Field Orientation, Turning Procedure

Let the shark, after coasting rectilinearly long enough to reset its electroreceptors, change the direction of its rectilinear path of travel by turning left or right at constant forward speed within a time short compared to the adaptation time constant of the ampullary sense organs, cf. Fig. 1.1d–g. Then, in the animal's frame of reference, a horizontal uniform *electric* field appears normal to the field in the seawater and rotating in the opposite direction of the animal's turning. This becomes self-evident by realizing that the electrical stimulus is given by the rapidly increasing vector difference between the field at the beginning of the turn and during the turn. The initial strength and polarity of the signal is indicated in Fig. 1.1f, g by the heavy and light + and − signs. Note that the initial equipotential parting line between the positive and negative regions of the ampullary system indicates the direction of the ambient electric field. Hence, one single turn suffices to determine on short notice the strength and direction of the field. See also Fig. 1.3. Once more, I must warn the reader that it may not be easy to follow the reasoning and require repeated contemplation.

Upon close inspection, it becomes evident that the electric field the animal detects rotates in the animal frame in the opposite direction over only half the angle the animal turns in the earth frame. This is readily verified by drawing a vector diagram of the field before and after the turn and considering the rotation of the vector difference between the two, see Fig. 1.4. Hence, the parting line between the positive and negative regions in the horizontal plane of the animal's head initially points in the direction of the ambient-electric field, but then starts lagging by half

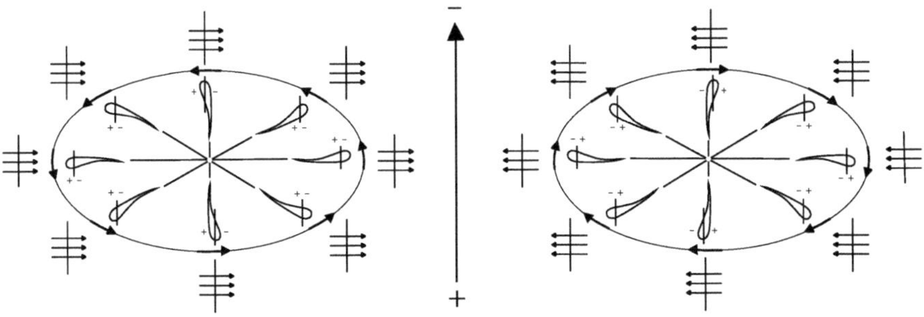

Fig. 1.3 Shark turning in uniform ambient electric field after resetting ampullary sense organs by moving rectilinearly for several seconds. Left: shark turning counter-clockwise. Right: shark turning clockwise. Large arrow in the center indicates the direction of horizontal, uniform ambient electric field. Miniatures depict stimulus field (three parallel lines with arrows) the animal receives from turning and cross-section of one equipotential plane (single line normal to three field lines), which like straight lines across the head of the shark indicates parting line between pore + and pore – regions. Note that the initial parting line, separating the + and – regions, gives the animal direction of ambient field. Electric fields are those received at the onset of turn

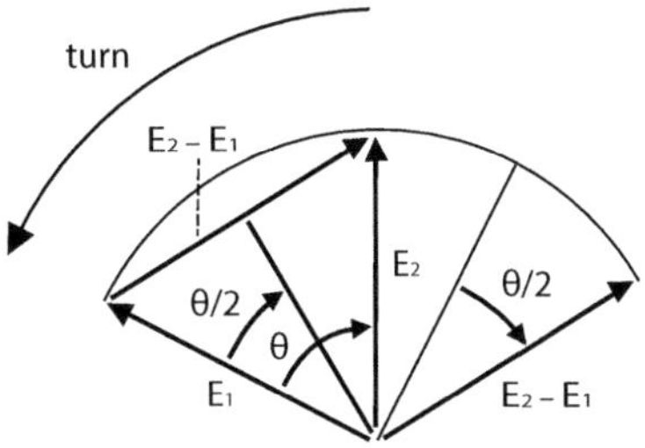

Fig. 1.4 Vector diagram. Animal observed in ocean frame. $\mathbf{E}_1$ electric field received by the animal before left turn, $\mathbf{E}_2$ electric field received by the animal after left turn, $\mathbf{E}_2 - \mathbf{E}_1$ difference signal detected by the animal. Note that the signal detected in the animal frame turns in the opposite direction over half the angle the animal turns in the ocean frame for turns less than 90 degrees

the turn. The strength of the stimulus is proportional to the strength of the ambient electric field and, initially, proportional to the angle of turning, but later on, more precisely, proportional to the length of the cord subtending the angle of turning. As corollary, from Fig. 1.4 it is obvious that, when a shark makes a constant-velocity U-turn within a time period short compared to the time constant of sensory adaptation, the difference in the field it detects just before and immediately after the U-turn is twice the strength of the actual ambient electric field and opposite in direction. This offers the animal a quick method to take a snapshot of the existing ambient electric field.

Conceptually, we may arrive at the same conclusion by realizing that the ampullary sense organs having their canals along the direction of the ambient electric field receive the weakest signals at the start of the turn, thus give the direction of the ambient electric field, whereas the organs having their canals normal to the ambient electric field receive the strongest signals at the start of the turn, thus give the strength of the ambient electric field. It could not be simpler.

But how might the animal keep the signals received from the ambient electric field apart from the signals received from moving in the earth's magnetic field? *When the animal turns left or right at constant forward speed, it receives a clear signal from the horizontal component of the ambient electric field without receiving an interfering electrical signal from interaction with the earth's magnetic field, neither from the horizontal nor the vertical component. When the animal turns up or down at constant forward speed, it receives a clear signal from the vertical component of the ambient electric field without receiving an interfering electrical signal from interaction with the earth's magnetic field, neither from the vertical nor the horizontal component. Hence, sensing the ambient electric field in this manner does not depend on the presence of the earth's magnetic field at all.* Since the animal receives no signal from the vertical component of the ambient electric field when turning horizontally at constant forward speed but does receive an electrical signal from interaction with the horizontal component of the magnetic field along its ventro-dorsal axis, it receives concurrent information about its magnetic compass heading. Similarly, by moving up and down at constant forward speed, it receives information about the vertical component of the earth's magnetic field from the electrical signals it received along its left-right axis. One caveat mentioned earlier is the observation that during turns, the left and right side of the animal slightly change in speed. However, their interaction with the earth's magnetic field causes opposite electric signals on opposite sides of the body canceling each other differentially. Certainly, evolution excellently solved a tough electrical engineering task.

How strong are the electrical signals a shark receives from turning through an angle of 30 degrees in uniform ambient electric fields? When, after resetting the sense organs by adaptation, the animal turns through an angle of 30 degrees within a time short compared to the adaptation time constant, the signals the ampullary canals directed normal to the electric fields receive in ambient electric fields ranging from 50 to 500 nV/cm is about 50 cos (30) = 50 times 0.866 nV/cm to 500 cos (30) = 500 times 0.866 nV/cm. Actually, the animal need not reset the sense organs, but merely needs to detect the change in the signal from the beginning to the end of the turn. Hence, during the animal's normal behavior, the method of detecting the directions of ambient electric fields by turning is effective over its full range of headings.

3.5 Magnetic Compass Orientation, Turning Procedure

So far, we have considered two rectilinear equal-speed exploratory moves in different magnetic compass directions. Let us now consider the two individual rectilinear moves performed in a single compound move by connecting them by a single left or right turn, the whole compound move to be performed at constant forward speed within a time short compared to the adaptation time of the sense organs. Besides during the rectilinear stretches not being affected by the ambient electric field, neither with the horizontal nor with the vertical component, the ventro-dorsal signals received by interaction with the horizontal component of the earth's magnetic field before and after turning left or right at constant forward speed constitute an equivalent, continuous dual-move, in addition to the individual two-move procedure for giving the shark its magnetic compass heading and a measure of the strength of the horizontal component of the earth's magnetic field. Moreover, the left-right signal received during the rectilinear parts of the compound move as well as during the turn gives the animal a measure of the strength of the vertical component of the earth's magnetic field.

Be aware, however, it is not the turn per se *that generates the signal providing the directional information. Instead, it is the change in the direction of the animal's rectilinear motion that leads to the biologically adequate ventro-dorsal stimulus. Actually, the turns* per se *cause circular currents within the animal's body that, by anatomical design, do not effectively stimulate the ampullary sense organs. Hence, the term turning procedure is somewhat of a misnomer, as it differs only from the individual rectilinear moves discussed before in that the two successive moves in different directions are smoothly merged together to form one smooth compound move.*

For small turns, the ventro-dorsal signals are initially proportional to the *change* in the sine, hence to the cosine, of the angle θ made with the horizontal component of the earth's magnetic field. A brief inspection reveals that the strength of the signal is in four directions on the magnetic compass rose the same, Fig. 1.3. Taking the polarity of the signal into account, this reduces to two directions. Taking furthermore the sign of the derivative into account, that is, whether the signal during the turn progressively increases or decreases in strength, this reduces to one unique direction. The method may be applied to two single moves or one compound move in different directions, or by repeated changes in the direction of travel, as during the shark's meandering mode of swimming. The latter, though, has the drawbacks that (1) the cruising speed of quietly swimming sharks is rather low compared to the potentially high speeds reached in single spurts, and (2) the swerving arcs are rather small compared to those seemingly erratically single moves sharks normally make. Nevertheless, the physical principles are very much the same. For further detail, see Fig. 1.5 and its descriptive legends.

How strong are the electrical signals a shark receives from turning through an angle of 45 degrees by interacting with the horizontal component of the earth's magnetic field? When, after resetting the sense organs, the animal changes its compass heading by 45 degrees at constant forward speed within a time short compared to the adaptation time constant, the electrical signal it receives at a forward speed of 1 m/s in a horizontal magnetic field of 40 μT, varies as a sine function, ranging from about 160 nV/cm along the E-W axis to 16 nV/cm along the N-S axis. Hence, the method of turning by 45 degrees is effective over the full range of compass headings.

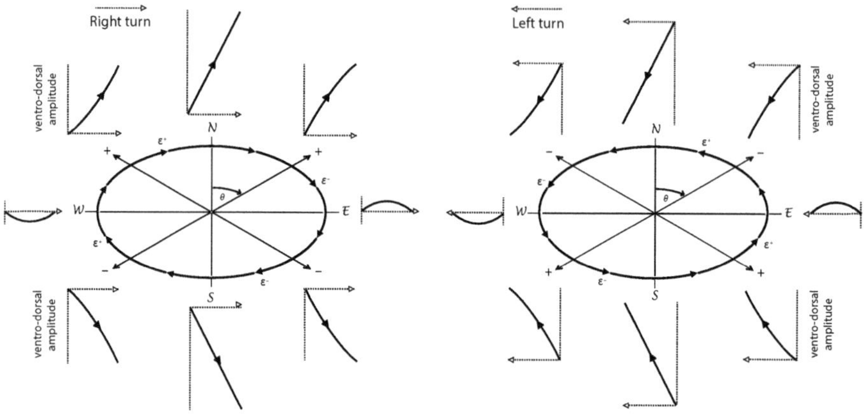

Fig. 1.5 Shark exploring magnetic compass heading *by turning right or left at reproducible constant forward speed* through 45-degree angles centered on eight different compass directions according to arrows strung along ellipse. The signatures are plotted as a function of the angle turned, positive angle to the right, negative angle to the left. The strength of the ventro-dorsal signal, which for small turns is proportional to the change in the sine, that is to the cosine of the angle θ, is the same in four directions on the magnetic compass rose, say, NE, SE, SW, NW. By taking the sign of the signal into account, that is, whether the signal is ventro-dorsal (+) or dorso-ventral (−), this is reduced to only two directions, NE or NW and SE or SW, respectively. The selection between these pairs of directions may be made by considering the change ε in the slope of the signal during turning into an adjacent direction, which depends on the + or − sign of the derivative of cosine, that is, on the negative of the sign of sine θ. Thus, when the sign of the negative of the sine θ is positive, the signal progressively increases or decreases faster when the turn is to the right and slower when the turn is to the left, as in the SW and NW directions, and, *vice versa,* when the negative of the sign of sine θ is negative the signal progressively increases slower when the turn is to the right and faster when the turn is to the left, as in the NE and SE directions. Hence, taking the progressively faster or slower increase or decrease of the signal during the right or left turns into account, unambiguously reduces the selection to just one, unique compass heading. Let E_{rel} be the strength of the signal divided by the strength of the signal when the animal makes a similar turn through north, then, expressed logically, the angle θ is given by ($\theta = \cos^{-1} E_{rel}$ AND $\varepsilon > 0$) OR ($\theta = \sin^{-1} E_{rel}$ AND $\varepsilon < 0$). When crossing the N or S directions, the signal is strongest and the sign of the sine changes. When crossing the W or E directions, the signal is weakest, and the sign of the angle remains the same. For larger turns, the signal may be calculated to any desired precision by developing the cosine of θ in a Taylor series: $\cos(\theta + \delta) = \cos\theta + 1/2\,\delta \sin\theta + 1/6\,\delta^2 \cos\theta + \ldots$ Note that with the signals the shark receives, we mean the signals in response to its exploratory movements, its turning at constant forward speed, those adding to any pre-existing excitations. The signatures are plotted as a function of the angle turned. If the speed of turning is not linear, the shape of the signatures will change accordingly. Note: Turning in the horizontal plane does not cause an effect from interaction with the vertical component of the earth's magnetic field. However, it does from interaction with the horizontal component of the earth's magnetic field inducing closed loop currents internal within the animal, which the shark's ampullae of Lorenzini are not sensitive to by design, as their skin pores open to the outside seawater

3.6 To Summarize the Various Exploratory Movements

Shark exploring earth's magnetic field and horizontal ambient electric field.

Rectilinear motions: shark accelerating and decelerating in horizontal plane:

Ventro-dorsal response to horizontal component of the earth's magnetic field.
Left-right response to vertical component of the earth's magnetic field.
No response to ambient electric field.

Angular motions: turning left or right at constant forward speed in horizontal plane:

Horizontal response to horizontal component of ambient electric field.
Vertical response to horizontal component of the earth's magnetic field.
No response to vertical component of the earth's magnetic field.

Actually, for a shark to detect the ambient electric field as well as its compass heading, it may combine the various individual exploratory movements to form a highly effective sequence. Let the animal start from rest after letting the sense organs reset. First move: let the animal rectilinearly accelerate to a certain velocity in an arbitrary compass direction. Second move: let the animal turn right or left through an angle of, say, 30 degrees while maintaining a constant forward speed. Third move: let the animal rectilinearly decelerate in the new compass direction to come to rest again. Let this all take place well within the time constant of sensory adaptation. During the first and third moves, the animal receives the information along its ventro-dorsal axis needed to determine its compass headings, while during the second move the animal receives in the plane normal to its ventro-dorsal axis the information needed to sense the horizontal component of the ambient electric field. Most importantly, the information will be available on very short notice, basically all the time, as required during electrically evoked predatory responses.

3.7 Ambient-Electric and Compass Orientation, a Worked-Out Example

From the intricate reasoning thus far, we have learned that, when a shark travels in a horizontal plane, a common occurrence, it can unambiguously discriminate between the ambient-electric field and the electric field giving the animal its magnetic compass heading and magnetic latitude on the earth's globe. Linear accelerations in two different directions suffice for the shark to determine its magnetic compass heading without being affected by an ambient-electric field, while single left or right turns not only give the animal its magnetic compass heading, but also inform it of the direction.

Figure 1.6 reviews the electrical signals received by a shark tracing an octagonal circuit. The miniatures and signatures each belong to single moves executed sequentially. Each move starts with the sense organs reset to zero, so that the results are

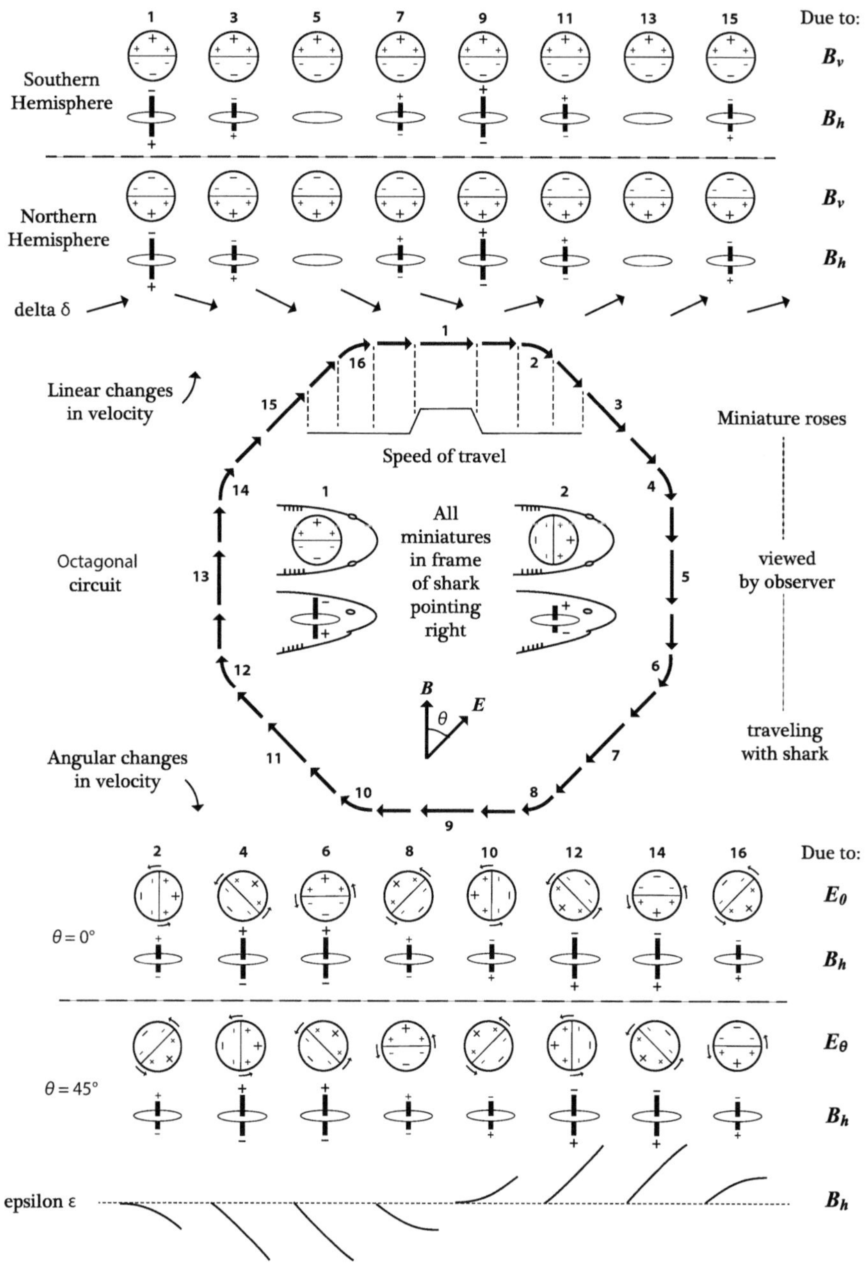

Fig. 1.6 Electrical signals detected by a shark tracing an octagonal circuit in the presence of the earth's magnetic field and a uniform, ambient-electric field. Lengths of straight and curved arrows along octagonal circuit indicate shark's speed of travel. Odd numbers refer to parts of straight stretches along which shark accelerates and decelerates, even numbers refer to curved parts of path along which shark changes its heading without changing its speed. For each of the numbered parts

(continued)

of the track, the horizontal electric signal the animal detects is indicated by a miniature circular rose (top view), and the vertical electric signal by a miniature oval rose (side view, slightly from above). The roses are considered fixed in the head and, hence, to rotate with the animal. The shark is depicted facing to the right in its own frame of reference, in which it is not the animal but the world that accelerates or turns. The horizontal component $\mathbf{B}_h$ of the magnetic field points to the top of the figure. The horizontal component $\mathbf{E}_h$ of the electric field makes an angle θ of null or 45 degrees with the horizontal component $\mathbf{B}_h$ of the magnetic field. Top row miniatures: transient electric signal shark detects along straight parts of the circuit by rectilinearly accelerating and decelerating in the *earth's magnetic field* in southern and in northern hemisphere. See also situation #1 inside the octagon. The left-right signal is independent of the direction of travel, pointing from left positive (+) to right negative voltage (−) in the northern and in the opposite direction in the southern hemisphere. The vertical, ventro-dorsal signal informs the animal of its magnetic compass heading. The differences between 3 and 15 and between 7 and 11 is given by the increase or decrease of the signal after the turn, as indicated by the slanted arrows delta δ. The ambient electric field inherently does not affect the sense organs during rectilinear accelerations and decelerations. Bottom row miniatures: transient electric signal shark detects at the start of each 45-degree turn, executed without changing forward speed, for two directions of the *uniform ambient electric* field, $\theta = 0°$ and $\theta = 45°$, referenced to north. See also situation #2 inside the octagon. The small, curved arrows on the perimeter of the horizontal roses indicate the apparent rotation of the parting line between the positive and negative regions during each of the turns from start to finish. At the start, the parting line points in the direction of the *ambient electric field*. After that, the parting line rotates with respect to the head through half the angle shark turns with respect to the water, though in the opposite direction. Yet, the next turn starts at the full angle of the previous rotation. The vertical component of the ambient electric field inherently does not affect the sense organs during left-right turns. The vertical, ventro-dorsal signal informs the animal of its *compass heading*, as its initial strength is proportional to the sine of the angle between the forward direction and the horizontal component of the earth's magnetic field. The differences between 4 and 6 and between 12 and 14 are given by the curvature of the signatures epsilon ε. The vertical component of the earth's magnetic field inherently does not affect the horizontal left-right signal at constant forward speed. Where the ambient electric field signals are less conspicuous either larger individual turns or several turns in succession, as during steady swimming, may be made to resolve the issue

those obtained from the move itself. The moves are considered short enough in time to consider the ampullae DC-sensitive during the various steps. To get a continuous record, the overlapping moves may be strung together, taking the differentiating time constant into account. The present presentation shows that changes in forward speed in two different directions and single turns left or right suffice for a shark unambiguously to detect the direction and strength of an ambient electric field as well its magnetic compass heading.

3.8 Swerving Motions During Steady Swimming

From the outset, I have reasoned that (1) to orient with reference to the horizontal component of the ambient *electric* field, left and right turns at constant forward speed are the most effective and readily understood exploratory movements sharks may make, where the constant forward speed is stipulated to prevent interference from interaction with the vertical component of the earth's magnetic field, and (2) to orient with reference to the horizontal component of the earth's *magnetic* field, two reproducible rectilinear accelerations in different compass directions are the most effective and readily understood exploratory movements sharks may make, where the accelerations are stipulated to be reproducible and rectilinear so as to enable the animal to compare the results obtained from exploring two different directions. The two individual rectilinear accelerations in different compass directions may also be executed as a single compound move, starting with the rectilinear acceleration, followed by a left or right turn into a different compass direction and, still at constant forward speed, proceeding rectilinearly again. The moves are all assumed to be fast enough to take place well within the time of adaptation to the DC content of the motional electric fields received, that is, by making excellent use of their low-frequency AC ampullary sense organs to detect the—at constant forward velocity—essentially DC motional electric signals. These simple moves may tentatively be considered elements of more complicated exploratory behaviors. Yet, it is not always that elementary. The two procedures, the individual and compound moves reviewed here, were only meant to demonstrate the physical and biological feasibility of the shark's electric and magnetic orientation in its simplest form.

That the moves discussed thus far are most educational for us does not mean that the sharks and rays rely on them solely. Certainly, the animals' seemingly erratic behavior of randomly turning left or turning right, accelerating and decelerating, is fully consistent with the envisioned exploratory movements for orienting in the ambient electric and magnetic fields. After having explained that in my Kalmijn (1974) handbook chapter and subsequent articles, I also foresaw that even the sharks' meandering mode of swimming may suffice to monitor both the ambient electric as well as the earth's magnetic field, although the rectilinear and turning moves take place concurrently. However, during steady travel the angles of turning remain rather small, and the DC drops out, except that the varying thrust of the tail-beat renders the speed of travel slightly rippled. Hence, I did not propose the traveling behavior in preference to the individual exploratory moves, but I proposed it as a possible extension of the theory, despite its limitations. In contrast, the individual moves produce stronger signals and are more effective to inform the animal of the ambient electric and magnetic fields. Yet, I expect the animals to make good use of all different moves, simple, compound, and concatenated.

Parenthetically, we may wonder how sharks and rays operate their electric and magnetic sense under different circumstances? We know, they use their electric sense in predation, guided to their target by the bioelectric fields of their prey. However, their fast-guided approaches are executed in the presence of the ambient electric and the earth's magnetic fields. Hence, the omnipresent electric and magnetic fields they rely on in open-ocean orientation must in predation be taken into account momentarily. That means, sharks and rays must all the time be fully informed of the strengths and directions of the local ambient electric and magnetic fields and correct for the signals they receive from interaction with them, unless the guided approach can be made independent of them. I have thought of it, but simply do not know the answer, yet, if I ever will.

The reason for bringing up the issue is that, for sharks, it does not suffice to orient electrically and magnetically during their long-distance travels only upon reaching a steady swimming routine, after a while eliminating the all-important DC contents of the electrical stimuli they receive. It makes more sense, to me at least, that the animals initiate their magnetically oriented directed travels by first performing the proper acceleration and turning moves so as to start out in the desired magnetic direction and then, while swimming steadily, simply maintain that direction, which does not require an elaborate processing, only keeping the stimulus pattern received from interaction with the ambient electric and magnetic fields constant, whatever that stimulus pattern may be. Perhaps, some scientists expect too much from the central processing during travel. To be sure, the animal may now and then execute accelerations and turns to verify their intended compass heading.

During steady travel, the signal received from turning left and right in the ambient *electric field* is inherently of an AC nature, independent of the forward speed, including turning on the spot. However, the signal received from merely turning in the *magnetic field* without forward speed does not produce an effective stimulus by the anatomical layout of the ampullary sense organs. The left and right turns are only needed to change the direction in which the head is moving with respect to the direction of the ambient magnetic field. At the start of travel, the animal's acceleration produces large motion-electric signals. Yet, after the strong, informative DC content of the stimulus resulting from the initial acceleration along the path of travel has died out due to the DC adaptation of the sense organs, what signals are left for the shark to orient electrically and magnetically? There are currently two alternative models addressing this question under discussion, one abstruse and inadequate, the other transparent and effective, respectively, labeled A and labeled B.

To orient in reference to the horizontal component of the ambient *electric* field, the swerving mode of swimming is excellently suited. As explained above, the direction in which the horizontally oriented ampullary sense organs are, during the swerving mode of swimming, stimulated least gives the direction of the ambient electric field, while the ampullary sense organs that are oriented normal to the direction of the field give the strength of the ambient electric field. During normal swimming the turning motions all take place well within the time constant of adaptation. At constant speed in the direction of travel, the ventro-dorsal signals due to interaction with the ambient magnetic field do not interfere with the detection of the horizontal component of the ambient electric field, as explained earlier. In that respect, orientation in the ambient electric field is very effective and robust. Actually, the detection of the ambient electric field may take preference over the detection of the earth's magnetic field as I have experimentally observed, as mentioned elsewhere in the present article.

To analyze the situation of a shark steadily swimming in the earth's *magnetic* field, we may, to make the analysis tractable, (A) assume the animal to explore the magnetic field by sinusoidally changing the direction in which the head moves, while keeping the speed of the head in that direction constant. That would just be a logical extension of the procedures I have contemplated so far: turning at constant forward speed. Or (B), we may represent the motion in the direction the head moves by the vector sum of (1) the head's sinusoidally varying motion *normal* to the path of travel in recoil of the tail beat, causing in the reference frame of the animals varying rectilinear accelerations and decelerations, and (2) the varying motion of the head *parallel* to the path of travel as a result of the shark's acceleration to get up to speed and the ensuing varying thrust during steady travel. The latter procedure is mathematically valid because of the distributive nature of the vector product. Let us discuss both models, (A) and (B), to ascertain which one is most realistic.

(A) If we assume the speed in the direction the head moves to remain constant but the angle between the direction the head moves and the direction the animal travels to vary sinusoidally, the analysis becomes mathematically unnecessarily complicated, as it is not a true to life assumption. If the direction in which the head moves is varying sinusoidally about the direction of travel and the speed in the direction of travel is constant, the speed in the direction the head moves cannot remain constant, and conversely. Moreover, the forward thrust of the tail renders the forward speed along the direction of travel not truly constant either. Hence, for those two reasons, the unrealistic assumption of the constant speed of the head in the direction it is moving leads to unrealistic results. This model and its reception in the literature will be discussed further in chapter 4.

(B) If we assume the head of the shark to move sinusoidally normal to the path of travel, then, at constant forward speed in the direction of travel, the ventro-dorsal signal is sinusoidal as well. However, the varying thrust causes the speed in the direction of travel also to fluctuate, adding an additional signal along the shark's ventro-dorsal axis. Both the motions normal to the path of travel and the motions along the path of travel are in the literature considered approximately sinusoidal, but they interact with the magnetic field in 90-degree different directions. By adding the signals from the two components, the analysis becomes mathematically rather simple. The ventro-dorsal signal then consists of a large ventro-dorsal sinusoidal signal due to sideways swerving plus a much smaller ventro-dorsal signal of different phase, due to the varying thrust, rendering the model very realistic.

The question remains, which of the signals gives the animal its magnetic compass heading, the ones due to the assumed sinusoidally varying motions *normal* to the path of travel or the ones due to the assumed sinusoidally varying motions *along* the path of travel? Both could serve equally well, but during steady swimming the variations along the path of travel are much smaller than the variations along the path normal to the path of travel, hence the motions normal to the path of travel take priority. Nonetheless, the variations in the forward motions may provide additional information, especially when the animal accelerates or decelerates rapidly. The information they provide is the same as that obtained by single accelerations in two different directions or by the compound moves discussed above, although the latter provide much stronger signals in general.

Two consecutive rectilinearly accelerated moves or one rectilinearly accelerated move and a single right or left turn at constant forward speed can be executed leisurely up to speeds of 0.5 m/s or in spurts up to speeds of 5 m/s. At a horizonal earth's magnetic field strength of 20 μT, the motionally induced signals then range from 100 nV/cm to as high as 1000 nV/cm due east or west with a sinusoidal fall-off to zero toward due north or south. Quite to the contrary, during steady travel, regular swerving motions over arcs of 22.5 degrees produce signals ranging from 40 nV/cm due north or south to merely 2 nV/cm due east or west. Signals of 40 nV/cm are still readily detectable, but signals of 2 nV/cm marginally so, but the deviations from purely sinusoidal are hardly detectable if not completely buried in the noise.

Actually, during travel, the animals are not restrained to swim steadily. Moreover, steadily still does not mean that their forward speeds are constant, as they depend on the periodic thrust of the tailbeat. Actually, sharks are known to change their travel, intermittently varying frequency, width of swerving, and direction, left and right, even up and down. By doubling their speed, the animals may not only reintroduce part of the DC component due to their forward motion, but also make the signals they receive four times larger, two times by doubling the frequency of swerving and two times by doubling the width of swerving. By tripling their speed, the signal will even increase some nine times. That would be worth the effort. In addition, the animals may change the direction of travel to explore the magnetic field even more effectively.

Having said the above, yes, I feel confident that the meandering mode of swimming assists sharks in maintaining a desired compass heading, not because there is no better or more readily understood means of doing so, but because I strongly believe the animals use anything that is detectable and provides vital information. Redundancy is the hallmark of safety. Any response pattern that is unique for the various compass headings may be of use, at least, as long as the signal is sufficiently strong for the details to be detectable against the noise of the sensory system. We may not logically understand how the complicated waveform comes to pass, but the animal may, so to speak, have an experience library of waveforms to consult. The animals know how to orient magnetically. We merely try to understand their abilities from a physical and logical point of view. Do not deny the animals their skills just because we have trouble explaining them.

In contrast, the signals received from the ambient *electric* field during the shark's meandering mode of swimming are readily detectable and equally strong in all orientations of the animal with respect to the ambient electric field. As explained above, it is plainly the initial parting line between the positively and negatively stimulated receptor pore regions that indicates the direction and polarity of the ambient field. *That leads to the interesting observation that the animal may make a few large exploratory movements to determine its magnetic compass heading and then maintain that compass heading while traveling quietly by keeping a constant angle with the ambient electric field.* The animal apparently also keeps a constant angle with respect to the bioelectric fields of their prey to reach their target as I have reasoned and supported with behavioral observations.

How strong are the ambient electrical signals the quietly traveling shark receives? When in an ambient electric field, the head sways through a realistic arc of, say, 22.5 degrees, the peak-to-peak strength of the signal is about ¼ of the strength of the imposed electric field, *independent of the orientation of the animal with respect to the electric fields.* Hence, the animals are sensitive enough to maintain a constant heading with respect to the ambient electric fields in most ocean waters, as the ambient electric fields range from 50 nV/cm to well over 500 nV/cm. Actually, I observed sharks and rays rectilinearly crossing Bimini Bay turn first left and then right when the ambient electric field was instrumentally rotated left for 60 degrees and then relaxed back to normal, at a field strength on the order of 20 nV/cm.

3.9 Motion-Electric Field from Undulating Motions

Rays and skates propel themselves by three different methods. They may use the thrust of the tailbeat to swim like sharks, they may use their wing-like pectoral fins to create undulating waves traveling over the body in variable patterns, or use their pelvic fins to walk on the ocean bottom. To orient in the ambient electric and the earth's magnetic field, the animals could use the same turning and accelerating exploratory moves that sharks I assume rely upon. Yet, the undulating wave of the broad wing-like pectoral fins is of particular interest here, as I have behavioral evidence suggesting that they use those when orienting in the earth's magnetic field, often when remaining still on the spot.

When a traveling ray interacts with the horizontal component of the earth's magnetic field, but now by undulating its pectoral fins, the vertical motions of the fins and the water directly under and above the fins cause a pattern of motionally induced electric currents to travel horizontally across the body. The direction of these currents and the voltage gradients they produce will be normal to both the vertical and the horizontal components of the earth's magnetic field. Hence, they may indicate the animal's compass heading. Admittedly, I have endeavored to visualize the idea but must defer the issue to future analyses, by me or my successors, to determine the validity of my thoughts.

In behavioral experiments in which stingrays were trained to enter an enclosure in the magnetic west, but not in the magnetic east, or vice versa, the animals often hesitated at the entrance of the enclosures before deciding either to enter and to get a food reward or not to enter to avoid gently being driven out without food reward. At those moments of hesitation, the animal commonly made frantic undulating waves along the pectoral fins without moving forward. This gave the observers the strong impression that these were the exploratory movements made to determine the direction of the ambient magnetic field before deciding whether to enter the enclosures or not.

3.10 Generalization of Electric and Magnetic Procedures

Rectilinear motions: shark accelerating and decelerating in vertical plane:

Ventro-dorsal response to horizontal component of the earth's magnetic field.
No response to vertical component of the earth's magnetic field.
No response to ambient electric field.

Angular motions: turning up or down at constant forward speed in vertical plane:

Vertical response to vertical component of ambient electric field.
Vertical response to horizontal component of the earth's magnetic field.
No response to vertical component of the earth's magnetic field.

To relax the strict, idealized conditions, when a shark probes the earth's *magnetic* field by linearly accelerating up to a certain speed *while ascending or descending*, in the presence of an arbitrarily directed ambient-electric field, by the same reasoning, the animal can detect the strength and direction of the earth's magnetic field during the constant velocity rectilinear part of the move without being affected by the ambient electric field. The animal still measures the signals received from interaction with the magnetic field *with respect to its body axes*, but now the ventro-dorsal body axis is no longer parallel to the gravitational vertical. Nevertheless, the component of the earth's magnetic field along its ventro-dorsal axis, rather than along the vertical, and the component of the earth's magnetic field in its frontal plane, given by the caudal-rostral and left-right axes, rather than in the horizontal plane, provide complete magnetic information. Hence, sharks may *in very general* by this method detect the direction of the magnetic field vectorially in three dimensions with respect to its own frame of reference, even when its own frame does not coincide with the earth frame, but the two frames are tilted by the animal's angle of inclination when ascending or descending.

To detect the vertical component of the ambient *electric* field, the same reasoning applies as for the horizontal component, except that the animal has to turn in the vertical plane normal to the horizonal component of the ambient electric field at constant forward speed, that is rostrum-up or rostrum-down, to detect the vertical electric field parallel to the sagittal plane, the plane defined by the caudal-rostral and the transverse axes. During the turn in the vertical plane, the normal horizontal component of the ambient-electric field does not affect the animal.

By turning rostrum-up or rostrum-down without changing its speed, the animal does not create a motional electric signal in its sagittal plane from interaction with the horizontal component of the earth's magnetic field along its caudal-rostral axis, neither is it disturbed by interaction with the horizontal component of the earth's magnetic field along its left-right body axis, as the induced electric fields on opposite sides of the body are opposite in direction and differentially cancel one another. Obviously, when the ambient-electric field direction is arbitrary, the animal may, by turning successively left or right and turning up or down, determine the direction of the field with respect to its body axes in three-dimensional space.

As for the *magnetic* orientation by turning left or right at constant forward speed, also by turning rostrum-up and rostrum-down it is the ventro-dorsal magnetic component and the magnetic component in the frontal plane that the animal detects, rather than the vertical and horizontal components of the earth's magnetic field. Thus, the restrictions of swimming in a horizontal plane in a horizontal ambient-electric field can all be lifted if we speak of the magnetic compass directions *with respect to the animal's body axes*, which the animal may transform to the components of the earth's magnetic field based on their three-dimensional relations between its body axes and the vertical of the earth's gravitation field.

One final generalization must be touched upon. The various exploratory modes so far considered in isolation may and usually do overlap during normal behavior. Moreover, requirements such as accelerating rectilinearly and turning at constant forward speed may be relaxed. It depends merely on what the animals may expect from their exploratory moves in order properly to interpret the electrical consequences of the various motions. They analyze the whole ensemble of electrical signals, of course not by reasoning, but by innate and gained experience. The present discourse only shows the protocols the animal *could* follow, even though they may not, just to prove the logical feasibility of sensing the ambient electric field as well the earth's magnetic field with one and the same sensory system. It would be extremely interesting to learn how these venerable animals do so neurophysiologically. Here, I have only investigated what sensory cues enter and what behavioral actions are taken by the central nervous system, from where on I hand my studies over to the more cerebral specialists, though they too should abide by the physics and biology of what goes in and what goes out.

How fast and to what extent the shark accelerates and decelerates and how fast and over what angle it turns right and left is under the animal's voluntary locomotor control, yet the animal may also monitor the intended motions electrically as well as by the use of its linear acceleration-sensitive otolith organs and the angular acceleration-sensitive semicircular canal organs of the inner ear. Even the lateral-line sense organs and vision may play a role. However, fast reflex moves may be executed without sensory monitoring, for instance, in escaping from a danger.

Since we have limited the discussion to the simplified version of the electric and magnetic orientation theory, ignoring the charge accumulations on the boundaries of the stream and on the skin of the animal, are the electric and magnetic exploration modes still valid when the animal or the stream are only lightly or not at all electrically loaded? It ultimately depends on whether and how much the electric fields of these charge accumulations add to or subtract in strength from the driving motionally induced electric field in the ampullary canals. Hence, they do not affect the logic of the reasoning behind the exploratory movements, but they determine how effective those movements are in leading to sensory excitation. Too weak of a signal is of course of little value.

3.11 Time-Domain or Frequency-Domain Processing of Sensory Data?

Thus far, we have first and foremost discussed the directional information sharks and rays receive from executing single or compound exploratory moves in the presence of an ambient electric and the earth's magnetic field, by turning right or left and by linearly accelerating or decelerating. That is, by responding to electrical signals that are not periodic and the animals process in the time domain. Yet, during steady travel, the single exploratory moves are concatenated in overlapping fashion, creating periodic electric signals owing to the sharks' characteristic head-swaying mode of swimming. Might the animals process these periodic signals in the frequency domain rather than in the time domain in order to sense the direction of the ambient electric as well as the earth's magnetic field? To answer this fundamental question, we must broaden our view and investigate the operation of the sharks' electric sense in situations where the mode of processing is self-evident.

During foraging, sharks heavily rely on their keen electric sense. When hungry, especially when odor-motivated, they impulsively respond to unexpected local changes in the direction and strength of the ambient electric field with sudden fast moves that take them to the sources of the fields, in the hope and expectation that they turn out to be prey. Their guided approaches to the sources of the fields are highly stereotypic, even though the signals that the sharks receive vary greatly, not only depending on the shape and orientation of the prey's bioelectric fields but also on the local strength and direction of the ambient electric field and the local strength and direction of the local earth's magnetic field. Moreover, the strengths of the electrical signals sharks receive from these various sources are, in general, roughly on the same order of magnitude. Since the prey fields have components along all three of the sharks' body axes, the signals the animals receive during the guided approaches from the prey are inevitably mingled with those they receive along the same body axes from existing ambient-electric fields and from moving in the earth's magnetic field.

One cannot escape the conclusion that sharks must take into account the electrical signals they receive from interaction with the ambient electric and the earth's magnetic fields when executing their electrically guided approaches. This also implies that sharks, however erratically they move about, at the start of their prey attacks, that is, at any time, must be informed of the strength and direction of the ambient electric and the earth's magnetic field with respect to their body axes. Since sharks prove to be astoundingly proficient in concurrently processing the ambient electric, ambient magnetic, and prey fields, fast and on short notice, they cannot but process the sensory data in the time domain. Hence, a fast, time-domain processing is not only needed for the detection of the prey fields *per se*, but also to render the electrically guided attacks possible in the presence of the ambient and electric fields. The only alternative solution would be that the animals act according to an approach algorithm inherently insensitive to the presence of the ambient electric and magnetic field, a fancy thought.

If the animals indeed have to correct for the electrical signals they receive from interaction with the ambient electric and the earth's magnetic fields during their predatory attacks, as I believe, they will need precise information about the electrical consequences of their rectilinear and angular changes in velocity during the electrically guided approaches to the sources of the fields. Might sharks for that purpose collect the information from the linear and angular acceleration-sensitive sense organs of the inner ear? Or might they merely anticipate the changes in velocity resulting from a controlled tail thrust and directional steering based on innate or former experience? Animal and man in general anticipate the results of their motions and monitor the results to correct for deviations from expectation. Often, however, when speed is of survival value, there is no time for monitoring and correction. That may well be the case at the last stages of sharks attacking their prey.

As we have reasoned above, to detect the ambient-electric and earth's magnetic fields, a couple of well-controlled exploratory moves suffice to obtain complete information. Sharks need not swim steadily for a period of time to sense the ambient-electric field and their magnetic compass heading. They can constantly monitor the ambient electric as well as the earth's magnetic fields during their seemingly erratic behavior of locally wandering about. While steadily swimming, the animal can evaluate each single period in the time domain and do so repeatedly to refine the results. In the time-domain mode, there is no need to make the periods similar or wait until the response becomes steady after the initial transient, as the frequency-domain mode requires. Anyhow, the time-domain approach is strictly needed for sharks to act fast in prey detection, not only in response to the prey field but also to the ambient electric and magnetic fields in order to correct for them. Hence, one wonders whether on long-distance migrations the frequency-domain approach offers significant advantages, however attractive the idea may seem. There is no need for sharks to operate in the frequency domain even on long-distance migrations and, additionally, this would require an *additional* processing procedure, entirely different from the time domain procedure.

The reader may wonder what then the great differences are between processing the sensory data in the time domain and processing the sensory data in the frequency domain. This is not just a semantic issue, but a question of vital importance for understanding how the brain functions. The time and frequency domain approaches require entirely different neurophysiological strategies. Biologically, the most important difference is that, in the time domain, the animal acts on the sensory data promptly as they come in, whether or not the signal is periodic, whereas in the frequency domain, the sensory data must be collected after the signal, when periodic, has become steady, that is after the initial transient and for sufficiently long time thereafter. Clearly, when the speed of response is critical, as in predation, operation in the time domain, not only concerning the prey field but also concerning the ambient-electric and the earth's magnetic field, is a must.

When a shark is swimming steadily, for instance during long-distance migration, the frequency-domain approach may still seem a reasonable possibility. However, there is a fundamental difference between (1) processing a quasi-periodic signal period-by-period in the time domain, treating the periods as individual events, and averaging the successive results, and (2) processing the entire signal in the frequency domain, collected over several periods after the steady state has been reached. An author using the terms frequency, harmonic, and phase, evidently has the frequency-domain approach in mind. Yet, the vital question is not only whether the frequency-domain method is a reasonable approach but, even more importantly, whether evolution has incited the animal to operate that way, even though in the frequency-domain mode sudden transient events escape attention, unless the animal concurrently operates in the time-domain mode as well.

3.12 Afterthoughts

The mathematically inclined scientist may wonder why we did not just write down the governing equations and arrive at the results by solving those by use of a math program. Well, they would not have given more insight than the reasoning we followed in the present paper. The equations would merely be a shorthand of our logical reasoning. They certainly would make the thought process more concise and more quantitative, but whether they would give the general reader more insight, the author doubts. Actually, they would tend to hide important detail to the non-initiated. At any rate, whichever method we follow, all are probably foreign to how the animals' central nervous system arrives at the desired behavioral responses. We may, and shall, approach the issues from all sides, but we will never reach the ever-receding scientific horizon, nor fully understand the senses of animal and man. By good fortune, medical science on our side of the horizon has made it possible for me to write this article by keeping me alive and in good spirits for so many years despite some close calls.

Although the kind of detailed reasoning we labored through must have been overly tedious for the reader, do not forget, the signals we receive from our inner-ear gravity and acceleration-sensitive otolith organs and our angular rotation-sensitive semicircular-canals organs, operating in our proper frames of reference, present very similar challenges, which in daily life we learn to deal with leisurely. Indeed, sensory performances, in general, turn out to be incredibly intricate when analyzed in detail. Clearly, if one wishes to gain clarity in sensory studies, one must start from the underlying physical principles in order meaningfully to interpret the neurophysiological and behavioral observations, as will become painfully evident in the next section, written in the hope and expectation that it may free the study of electric and magnetic orientation in sharks and rays from senseless human quarrels and will raise it to the high level of understanding expected from pure science. I hope my sincere lifetime effort will not be in vain and, God forbid, will not cause even more confusion.

References

Kalmijn AJ (1974) The detection of electric fields from inanimate and animate sources other than electric organs. In: Fessard A (ed) Handbook of sensory physiology, vol III/3. Springer, Berlin, pp 147–200

Kalmijn AJ (1981) Biophysics of geomagnetic field detection. IEEE Trans Magnetics 17:1113–1124

Kalmijn AJ (1997) Electric and near-field acoustic detection, a comparative study. Acta Physiol Stockholm 161(Suppl 638):25–38

Chapter 4
Sensible and Reprehensible Misunderstandings

Warning: sharks are not afraid to defend themselves ferociously, neither am I
This is not an all-inclusive review of the literature concerning the theory of electric and magnetic orientation in sharks and rays. Only the articles that significantly contributed to my motional electric theory and those that are squarely in disagreement with my ideas are discussed here. It certainly does not signify a lack of respect for those authors and articles not mentioned, neither does it tacitly agree with all that has been said on the topic in articles not referenced here.

What follows is time and again very painful for me. I feel bereft of the field of research that I pioneered half a century ago with the high expectation to share the joy of scientific discovery with my peers. After a long uphill battle, I have no peers anymore. The following section is my last attempt to separate the sense from the nonsense. In science, removing error is at least as important as generating new ideas and discovering new facts, if not even more important.

My task has been very difficult, because in all cases I first had to find out what and why the authors are saying, however wrong it may be. That was the most difficult and time-consuming part of my job. Then, I had to explain what is right and wrong, not what I believe to be right or wrong, but was it right or wrong based on well-established scientific knowledge and reasoning. I still keep critically reading my manuscript to catch myself on misinterpretations.

For a biologist to defend himself or herself by saying "do not blame me, I am not a physicist" and for a physicist to say "do not blame me, I am not a biologist" are false excuses for those studying the senses of animals, especially the senses of aquatic animals. If you do not understand the physics of a stimulus, how can you understand the response to it? If you do not know an animal's behavioral response, how can you speak of the biologically adequate stimulus? Either the research is conducted by a scientist with a dual background in biology and in physics, or by two scientists, one having a background in biology and the other in physics. Still, both must have rigorous, higher-education trainings.

A. J. Kalmijn, *Theory of Electric and Magnetic Orientation in Sharks and Rays Revisited*, https://doi.org/10.1007/978-3-031-58445-9_4

4.1 Before 1974, the Emergence of Electroreception in Sharks and Rays

Longing to admire, it is the task of scientists to find out

In deference to Faraday's 1832 tenet "*In the high and pure philosophic desire to remove error as well as discover truth,*" I consider it my duty as a scientist not only to advance what I consider true but also to rebut what I consider false. Of course, the criteria for discriminating between true and false must be consistent with well-established scientific knowledge and logical reasoning. If I fall short in that respect, I am wrong and must be corrected forthwith. However, in giving my opinion, I do not claim, but only believe to be correct, realizing that I may be mistaken regardless. I am well aware that scientific articles are written by scientists like myself and criticizing articles may hurt the authors. It does not make a truth-seeking peer beloved and invites revenge, as I too well know from experience. To assess my harsh comments, I advise the reader carefully to scrutinize the articles I find fault with and to consult with my earlier writings on the topic. If you disagree with me, please, let me personally know, so we can together seek to resolve the issues. It may be difficult to convince the original authors of their imputed failings, but perhaps you and I can together set the arguments straight. Yet, do not quote dubious articles without voicing your concerns in clear terms, as you will be liable for implicitly endorsing and perpetuating erroneous assertions and false claims.

Faraday, the founding father of magneto-electric induction, from wire loops to ships traveling at sea

Although Faraday did not foresee how sharks and rays orient in the vast open-ocean waters by his newly discovered principle of magneto-electric induction, he did infer in his Farady 1832. Bakerian Lecture that "*a ship or boat, passing on the surface of the water in northern or southern latitudes, should have currents of electricity running through it directly across the line of her motion.*" In that respect, I may be considered a student of Faraday, even though I was born 101 years after he put his visionary inference in writing. In my graduate years, I was captivated by his *Experimental Researches in Electricity*, which I read with great delight. They became the ultimate source of inspiration for my entire career in science. In studying the electric eel *Gymnotus*, Faraday (1839), as a true naturalist, exclaimed: "*Wonderful as are the laws and phenomena of electricity when made evident to us in inorganic and dead matter, their interest can bear scarcely any comparison with that which attaches to the same force when connected with the nervous system and with life.*"

First inklings of an electric sense, from weakly electric organs to ampullae of Lorenzini

The first reasoned conjecture of motionally induced electric fields affecting electrosensitive animals was advanced by Lissmann & Machin in 1958, who observed the weakly electric fish *Gymnarchus niloticus* behaviorally respond to a permanent magnet vertically swiped in front of the glass aquarium. They tacitly assumed the horizontal $\mathbf{v} \times \mathbf{B}$ at the position of the fish to be the physically adequate stimulus, instead of the ohmic voltage gradient $-\rho\mathbf{J}$ resulting from the electric current density field driven by the $\mathbf{v} \times \mathbf{B}$ field throughout the aquarium and shaped by the electric field of the charges accumulating on the walls, the bottom, and the water surface of the aquarium, and on the skin of the animal, a frequent persistently perpetuated mistake. Based on the misconceived physically adequate stimulus, they calculated the fish to detect voltage gradients as low as 0.03 μV/cm, at least a factor of five more sensitive than they later, in Machin & Lissmann 1960, proved it to be more tenable behavioral tests. Lissmann & Machin further speculated that *Gymnarchus* might be able to tell the difference between swimming north-south and east-west, using the motionally induced electric fields they create, unaware of the drawback of the fish living in very fresh water and not being as sensitive to low-frequency electric fields as they thought at the time. Even though these well-conceived conjectures stranded on a misunderstanding of the intricate physics involved, the novel pioneering researches of Lissmann & Machin signified an epoch-making breakthrough in the exciting evolving field of electroreception, introducing a rich new branch of scientific research.

Electrophysiologically investigating the mechanical sensitivity of the ampullae of Lorenzini by recording from their afferent nerves, Murray (1960a, b) noticed these, at the time puzzling sense organs, not only to respond to mechanical, in addition to thermal and salinity stimuli, but also to small electric currents he thought to have passed through the ampullary canals as well as to weak electric fields he introduced into the seawater directly over the mandibular ampullary organs. In the discussion of his 1960b article, Murray contemplated whether a shark might detect the strength and direction of tidal flows or ocean currents from the electric fields they induce by moving in the earth's magnetic field, but considered it improbable that the animal, being carried along with the water and thus subject to the same influences, could detect such weak fields, apparently not aware that it is just the tidal current or ocean stream moving with respect to deeper water or the ocean bottom that causes the $-\rho\mathbf{J}$ field the shark detects, as I became aware from the physical oceanographic literature in subsequent years. Though doubting about it, Murray had the right idea. Whereas he was not encouraged to delve deeper into the issue, at the time, I had just started studying the physics of the subject matter.

In a next article, Murray 1962 also contemplated the possibility of a shark detecting the vertically directed voltage gradient the animal induces by interaction with the horizontal component of the earth's magnetic field when swimming in an easterly or westerly direction. He calculates the motionally induced $\mathbf{v} \times \mathbf{B}$ to be 0.1 μV/cm/knot, the absolute value of which for sharks in ocean waters is not much different from the physically adequate stimulus $-\rho\mathbf{J}$ sharks actually detect. Since Murray electrophysiologically had arrived at a sensitivity of 1.0 μV/cm, he considered the electrical stimulus "to be about an order of magnitude too weak for the ampullae of Lorenzini to detect." Ironically, he concludes "although *Gymnarchus* could probably tell which way it was heading, it is unlikely that elasmobranchs can," though the opposite is more likely true, since the $-\rho\mathbf{J}$ *Gymnarchus* would detect is much smaller than the $\mathbf{v} \times \mathbf{B}$ in view of the light loading by the high-ohmic freshwater medium.

To Murray's defense, his peer reviewers did not object. In short, Murray, left it to me to resolve the riddle, for which he sent me his compliments. I wish, I would have had a chance to meet with Lissmann and Murray. I was too timid at the time to cross the English Channel. Yet, in the 1974 Springer Handbook on Electroreceptors, Murray's Chapter and mine are symbolically next to one another.

Dijkgraaf's long-ago observation and Kalmijn's doctoral thesis in physics and biology

My own life-time dedication to electroreception traces back to Parker and van Heusen 1917, who noticed the catfish *Amiurus nebulosus* to avoid or, at times, to approach and nibble at a metal rod, which the authors proved to be evoked by corrosion currents emanating from the metal surface, coming tantalizingly close to discovering an electric sense in fishes without electric organs. When Dijkgraaf, familiar with the article of Parker and van Heusen, in 1935 noticed sharks similarly to evade the metal wire he used to offer the animals food, he suspected this to be due to corrosion currents as well. When, more recently, Dijkgraaf learned of Lissmann's 1958 revealing discovery of the electric sense in weakly electric fishes, remembering his own earlier day observations he asked me to find out whether the avoidance responses of sharks indeed might have been electrically evoked and what the significance of the responses might be. That led to my doctoral thesis project. Hence, I measured the corrosion currents issuing from metallic objects with a DC preamplifier and AgCl electrodes I built as part of my concurrent physics project and tested the responses of sharks and rays to electric fields simulating the corrosion currents, with the result of proving Dijkgraaf right.

First, I noticed eyeblink and ventilatory reflexes in sharks and rays at the onset of uniform electric fields of 0.1 μV/cm, as I reported in Dijkgraaf and Kalmijn (1962). Then, recording the heartbeat in free-swimming sharks and rays, I found the animals to respond to fields as weak as 0.01 μV/cm, Kalmijn (1966), a hundred times weaker than Murray knew and ten times weaker than I previously had determined. That compelled me to search for the weak electric fields sharks and rays may detect in ocean waters. The first most conspicuous fields I found were the bioelectric ventilatory potentials that I measured in the vicinity of prey items. Soon, though, I became aware that the ventilatory potentials actually were low-frequency modulations of the prey's common DC bioelectric field, which I later observed in the vicinity of essentially all aquatic animals. Initially, I thought to have conditioned the sharks and rays to respond to the dipole fields electrically simulating prey, but after a while I found them also spontaneously to attack DC dipole sources, especially when the animals were aroused by the scent of injured fish *and the fields were weak enough*! These were tense times, as even Dr. Sven Dijkgraaf, my mentor, was skeptical about my wild ideas until I showed him my bioelectric field measurements and behavior tests in the presence of his trusted physics friend Dr. Jan Groen, highly respected for his pioneering research on the semicircular canals of elasmobranch fishes. Then Dijkgraaf became so excited that he published my graduate research with me, Dijkgraaf and Kalmijn (1966), upon which I wrote my Kalmijn (1966) Nature.

Move to California opens an ocean-wide opportunity
Before leaving my home country, The Netherlands, I worked on my Ph.D. thesis in an electromagnetically quiet underground World-War I bunker in the rustic vicinity of town. I conducted a year-long sequence of behavioral experiments conclusively *proving that sharks and rays detect and spontaneously zero in on the DC and low-frequency AC bioelectric fields of their prey, hence are not merely sensitive to electric fields, but have a true electric sense, now textbook knowledge*, see Kalmijn (1971), which—to my great delight—has been declared a "classic" in the Journal of Experimental Biology 2010 on instigation of Dr. Carl Hopkins 2010 at Cornell University, well known for his many years of research on the active, electric-organ mediated electric sense of weakly electric fishes, like myself, inspired by Lissmann's discovery. Yet, the credit goes to the sharks and rays that shared their secrets with me.

4.2 Definition of Electric and Magnetic Sense

Criteria for establishing an electric sense in sharks and rays, in my opinion
To declare that sharks and rays have an electric sense, we must not only prove that (1) the animals are adequately sensitive to detect the electric fields commonly occurring in their natural habitat but also that (2) they derive biologically meaningful information from the fields they detect. Whereas Parker and van Heusen speculated that catfish respond to the corrosion currents that metal rods produce when in contact with the aquarium water by inadvertently stimulating the taste buds, which in catfish occur widely spread over the skin, and whereas Dijkgraaf had the inkling that the ampullae of Lorenzini were galvanically stimulated by the metal wire he used to present small sharks with a piece of cut fish, neither of them advanced the idea that the responses might indicate that the animals have a genuine electric sense. The taste buds are for the sense of taste and the ampullae of Lorenzini were at the time considered mechanoreceptors, related to the lateral line. Hence, the responses to corrosion currents were considered *biologically inadequate* stimuli or, for short, *biological artifacts*, consistent with, but not proof of an electric sense. Neither the criteria of common natural occurrence nor of biological significance of the electric fields that the catfish, sharks, and rays responded to were met, however inspiring the observations might have been.

In the same year Murray (1962) contemplated the detection of the electric fields induced by ocean streams and tidal flows but concluded that the ampullae of Lorenzini were not sensitive enough, leaving the function of these mystifying sense organs undetermined. I behaviorally established the truly astounding electrical sensitivity of sharks and rays, inspired by Dijkgraaf's 1935 observation. However, when Dijkgraaf, my thesis mentor, in 1962 wanted to write an article with me as coauthor heralding our discovery that sharks and rays have an electric sense, I courteously but determinedly refused to go that far, as I did not want to speculate. I insisted that I first determine the naturally occurring electric fields sharks and rays may detect and what their *biological significance* might be, a term Dijkgraaf often used in criticizing his international peers studying the function of the lateral line and inner ear of fishes electrophysiologically. As part of my 4-year graduate biology and physics research, I succeeded in proving the *biological significance* of the astounding electrical sensitivity of sharks and rays, founded on (1) my measurements of the bioelectric fields in the vicinity of aquatic animals and (2) on my observation that sharks and rays spontaneously zero in on the bioelectric fields of their prey or purely electrical simulations thereof. Ultimately, in my 1971 article, I formally declared sharks and rays to have a genuine electric sense.

To recapitulate, by my strict criteria, sharks and rays have a genuine electric sense, since they unequivocally (1) zero in on the bioelectric fields of their prey, Kalmijn (1971) and—as I later found—(2) orient in the electric fields of ocean currents, Kalmijn (1982, 1988b), recounted below. These pioneering experiments have never been contested to my knowledge. Actually, the electrically evoked feeding response has been reproduced many times, though often in the form of biological artifacts, as the fields evoking the animals' responses were either not of bioelectric origin or did not properly simulate genuine bioelectric fields, such as the corrosion fields of iron nails hidden in the bottom sand or the motion-electric fields induced when the animals move through the field of a bar magnet buried in the sand or under the bottom of the observation tank. The attacks at rusty nails and magnets are, by my criteria, neither proof of an electric sense nor of a magnetic sense, since the target is not a commonly occurring natural object or a faithful simulation thereof and the elicited responses are not biologically meaningful. Hence, though suggestive consequences of the electric sense, they are, by my criteria, just biological artifacts that are improperly interpreted by the researchers.

Criteria for establishing a magnetic sense in sharks and rays

To declare that sharks and rays have a magnetic sense, we must not only prove that (1) the animals are adequately sensitive to detect the magnetic fields normally occurring in their natural habitat, but also that (2) the animals make biologically meaningful use of these fields in daily life. From my own experiments, I knew how sensitive sharks and rays electrically are. In Dijkgraaf and Kalmijn (1963), I proved together that the animals owe their electrical sensitivity to the ampullae of Lorenzini. We all know of the earth's magnetic field and of mariners using a compass needle to navigate the worldwide oceans. But how might the animals use their electric sense to detect the earth's magnetic field? Well, I learned that from Faraday (1832). To prepare myself for exploring the feasibility of the idea, I enrolled in the physics department to study electromagnetism side by side with my study in biology. That eventually led to my Springer Handbook Chapter of 1974 in which I introduced my theory of motional electric orientation, proving that sharks and rays receive all electrical information they need to orient in the electric fields of ocean currents as well as in the earth's magnetic field. Yet, I did not conclude to a true magnetic sense until I had proven that the animals behaviorally are indeed capable of orienting in the earth's magnetic field, as I had inferred on the basis of my motional electric theory. The definite "yes" answer came from my Kalmijn (1978b) experiments exhibiting the animals to orient in uniform magnetic fields of earth strength in a biologically meaningful manner.

In brief, sharks and rays have a genuine magnetic sense, as they *meaningfully* orient in the earth's magnetic field. However, behavioral responses to magnetic fields not *naturally occurring* in ocean waters and responses to magnetic fields of no biological relevance are not, in my opinion, proof of a magnetic sense. They only confirm that relative motion between a shark and the source of a magnetic field, such as a bar magnet held under the observation tank, induces electric fields within the sensitivity range of the animals' electric sense. If the animals respond to such fields anyhow by whatever means, the responses are, in my opinion, biological artifacts. For instance, sharks respond to the electric and magnetic fields of shipwrecks and to rusty shark observation cages, just as much as we respond to aluminum foil in our mouth if we have metal fillings or respond to the sparks given off when rubbing a woolen garment. On the other side, if we prove that animals have a genuine magnetic sense by my criteria, it is not strictly necessary to know how they detect the earth's magnetic field and which or even where the pertinent sense organs are, although not to know is singularly unsatisfactory and uncomfortably frustrating. Fortunately, in magnetic bacteria as well as in sharks and rays the detection and meaningful use of the earth's magnetic field is physically and biologically self-evident.

Difference between electric and magnetic sense in sharks and rays
Still, one may wonder what the difference is between an electric and a magnetic sense, if both rely on the electrical sensitivity of the ampullae of Lorenzini in sharks and rays. The crucial issue is that the ampullae of Lorenzini are the front-end sensors of a complete sensory system. The system as a whole, including the central processing and behavioral feedback, determines whether it acts as an electric sense or as a magnetic sense. The electrical signals received from one kind of exploratory movement are what we consider belonging to an electric sense. The electrical signals received by another kind of exploratory movement we consider belonging to a magnetic sense. The difference between the two modes is that the one conveys biologically meaningful information about—what we call—an electric field, such as the bioelectric fields of animals and the electric fields of ocean currents, and the other conveys biologically meaningful information about—what we call—magnetic fields, such as the ubiquitous earth's magnetic field. Remember, electric and magnetic fields themselves are two interwoven complementary features of one and the same electromagnetic field. Depending on the nature of the fields, the animals take the proper behavioral actions. Since sharks and rays receive biologically distinctly different information of what we call naturally occurring electric and magnetic fields, they have both an electric sense and a magnetic sense, by definition.

Here, I will make the confrontational statement that no magnetic experiments conducted to date in sharks and rays that I know of, other than my own, qualify by my criteria as proof of a true magnetic sense. I am aware that my experiments have been scrutinized, as they should be, but I have not heard or read anything that makes me less convinced that the results are less than irrefutable, not only that the animals have a genuine magnetic sense, but also that I have given compelling evidence that their magnetic sense is physically of a motional electric nature, mediated by the ampullae of Lorenzini. Other biologically meaningless reactions to biologically irrelevant magnetic fields, which I consider biological artifacts, may be a consequence of the animals having an electric sense and a magnetic sense, but do not prove either one, by my criteria. I know, this statement is bold, but below I will make clear why I dare say so, especially in reply to the many unfounded statements. In the present article, I endeavor to settle the issues once and for all, before it is too late, so as to safeguard my legacy and to advance science.

Incorrect suggestion of an additional criterion
Several authors add another restrictive criterion for speaking of a magnetic sense: in *their opinion*, the magnetic field must be detected "magnetically." This limiting criterion excludes the magnetic sense of sharks and rays from the magnetic navigation literature. These authors themselves, in general, wish that they knew how their own favorite study objects orient magnetically other than magneto-electrically, to use Faraday's terminology. I will come back to this issue below.

4.3 Sharks and Rays Practice Faraday's Physics

Introduction of physics governing electric and magnetic orientation in sharks and rays
After I had established the role of the electric sense of sharks and rays in the detection of prey, the question of orientation in the electric fields of ocean currents and in the earth's magnetic field became a challenging topic in dire need of a firm physical foundation. At this early stage, in 1974, I introduced the first version of my physical and biological theory of motionally induced electric and magnetic orientation, spurred by the conjectures of Lissmann & Machin and of Murray, and founded on the astounding electrical sensitivity of sharks and rays I had recently discovered. The initial version proved that, from a physics point of view, sharks and rays are not only day and night subject to electric fields presenting them with information about their drift in ocean streams and their magnetic compass headings, but also that the strengths of the electric fields they receive are well within the animals' sensitivity range I had determined behaviorally. At the time, though, I did not know yet whether sharks and rays actually make use of the electric and magnetic fields for the orientational purposes I envisioned, but the information available made it quite plausible that they indeed do so. Yet, without any earlier in-depth discussions on the topic, I first had to gain more insight as to how the animals might derive the desired directional information from the motionally induced electric fields before I could devise the proper behavioral experiments.

I still vividly remember my trepidation concerning the physical validity of my electric and magnetic theory of orientation in sharks and rays. Yet, I felt sufficiently confident to publish it, since I closely followed the thorough 1954 oceanographic article of Longuet-Higgins, Stern, and Stommel on "*The electric field induced by ocean currents and waves, with applications to the method of towed electrodes.*"

Yet, to be on the safe side, I left two issues for later clarification, afraid of unnecessarily confusing the reader or saying more than I myself was confident with. True, the two issues still seem to be stumbling blocks for many biologists trying to grasp the physics of motional electric fields and how they are detected by the animals. Just for that reason, it is important to highlight them in advance to prevent them from remaining nagging impediments to following my reasoning.

Firstly, I adhered to the mathematical conventions of the leading oceanographic scientists at the time in the field of motional electric induction, Longuet-Higgins, Stern, and Stommel, to call the *negative* of the voltage drop over distance, to wit, the *voltage gradient* $-\rho\mathbf{J}$, the physically adequate stimulus. For consistency, the authors expressed physical quantities in terms of gradients rather than forces. Indeed, electric field measuring devices, including the electroreceptive membranes of sharks and rays, measure voltage gradients, or rather the line integrals of them. As I explained in Kalmijn 1988a relativistic article, if one thinks of the forces on the gating charges of the voltage-sensitive ion channels, then it is the electric field one has in mind, which in the direction is the opposite of the voltage gradient, $\mathbf{E} = -\nabla\varphi$. In practical applications, one usually thinks of the integral term of the voltage φ and the differential term of the voltage gradient $\nabla\varphi$, but in the physical theory of electromagnetic fields, the term electric field $\mathbf{E}$ is more common. As long as one is aware of this, no harm is done.

Secondly, Longuet-Higgins *cs* 1954 did not explicitly mention the space and surface charges accumulating at the boundaries of the streams. Neither did I mention those, nor the charges accumulating on the skin of the animals that the motional-electric fields physically are associated with. I was aware that Longuet-Higgins made proper note of them in his earlier 1949 paper, but I knew by experience that many readers would be leery of space charges in volume conductors. Even though I specifically discussed those charges in Kalmijn 1988a and 1997 versions of the theory, showing them to be decisive and physically strictly requisite to arrive at a consistent system of stationary motion-electric current fields in ocean streams and traveling sharks.

Biological evidence proving feasibility of electric and magnetic orientation in sharks and rays

The original versions of my theory were based on logical inferences, starting from established facts. However, in Kalmijn (1977, 1978a, 1978b, 1981, and 1982) the theory got a strong biological foundation from my protracted series of electric and magnetic orientation experiments, proving that elasmobranch fishes, in particular the stingray *Urolophus halleri*, are indeed able to orient in the oceans' weak ambient electric fields as well as to sense their magnetic compass headings. In 1984, I extended the theory to include rotations in addition to translations. Below, I will brief the reader how various authors have tried unsuccessfully to undermine my behavioral experiments. My endeavor to envision the environment in which the animals live has led me to study their underwater habitat as perceived in the frame of reference in which the animal's sense organs are at rest. However, not all shark experts have been appreciative of my 1988a relativistic article. Special relativity is commonly thought only to be important at very high speeds, but is it really? Just read Einstein's "On the electrodynamics of moving bodies" (1905). He could just as well have spoken of sharks swimming in the ocean, instead of trains moving relative to the embarkment.

Prey detection all right, is it? But electric and magnetic orientation hard to believe

Interestingly, my observations on the electrically evoked feeding responses have never been challenged, although not all scientists have been able to reproduce them by lack of proper facilities or by inadequate animal care. True, it took me a long time to explain how sharks are guided by the bioelectric fields to their prey. But the more straightforward ambient electric and magnetic orientation of sharks and rays have met with reservations or even disbelieve. However, it has the beneficial effect that it forces me publicly to set the record straight in order to prevent upcoming scientists from believing that I tacitly agree with the many baseless, doubt-seeding remarks. Hence, with my rebuttal, I intend to give the reader the physical and biological insight needed to decide for himself or herself which of the articles represents the chaff and which represents the wheat.

Electrophysiological responses to biologically inadequate magnetic stimuli

Cognizant of the experiments of Parker and van Heusen (1917), Lissmann and Machin (1958), Murray (1962), and of my initial studies, yet before reading my 1974 Handbook Chapter, Andrianov et al. (1974) recorded the single-unit and evoked medullar responses to electric and magnetic stimuli applied to the ampullae of Lorenzini in the Black Sea skate, *Trygon pastinaca*. Their main objective was to analyze the functional characteristics of the central neurons, which they accomplished in a laudable effort. However, since the magnetic stimuli they applied did not simulate those skates received in nature when traveling in the earth's magnetic field, their results only demonstrate that changing magnetic fields induce electric fields that the animals can detect with the ampullae of Lorenzini. For the same physical reason, their comparison with a skate moving at a speed of 50 cm/s in the earth's magnetic field was mistaken. Whereas the magnetically induced stimuli they applied resulted in electric currents largely circulating within the body and within the seawater adjacent to the skin, the physically adequate electric currents a rectilinearly swimming skate induces traverse the body, leave through the skin, curve around through the seawater relative to which the animal moves, to return back through the skin again, naturally stimulating the ampullary sense organs.

Akoev et al. (1976) repeated the 1974 magnetic experiments of Andrianov et al. (1974), perpetuating the misconception of the physics involved in that the magnetic stimuli they applied induced electric fields that were fundamentally different from the motionally induced electric fields a skate receives when traveling in the earth's magnetic field. They, furthermore, stimulated individual ampullae with electric *currents* in the supposition that the current would be channeled through the ampullary canals. Murray (1962) originally thought so too, but in 1976 it was already long known not to be so, which the Russian authors were apparently not aware of. Hence, the current thresholds they report have little to no physiologically adequate meaning. The stimulus voltages they provide depend on the uncontrolled resistance of the surrounding tissues and seawater. Note: In later years, I have ascertained the ampullae of Lorenzini to be true voltage sensors with an extremely high, slightly negative input resistance, certainly not current sensors.

After reading my 1974 handbook chapter, they changed their mind and got it straight

At long last, after reading my 1974 Springer Handbook Chapter, Brown and Ilyinsky (1978) agreed with my account of the physics and, in addition to elaborating upon the earlier, biologically less meaningful experiments, they correctly recorded the nerve responses from the fish moving through the water and the water moving past the fish in the presence of a magnetic field. Their latter results proved to be in full support of my recent 1974 theory of electric and magnetic orientation. Kalmijn (1974) did not merely review the physics of electric and magnetic orientation by the principle of motional electric induction, as they say, but he was the first scientist to get the governing physics logically sound. Brown and Ilyinsky (1978) simply copied his equations, as if they themselves had just derived them from Faraday's law.

Motional-electric induction in animals not heavily loaded electrically by the surrounding medium?

Just like sharks and rays in ocean waters, terrestrial and airborne animals create, by interaction with the earth's magnetic field, motionally induced electric fields, but can they really detect them? When the animals *rectilinearly accelerate*, transient electric currents flow in circuits through the moving animal and the stationary air. They are merely transient currents, because in air the animals are only loaded capacitively. However, when the animals *turn* in the earth's magnetic field, steady closed-loop electric currents flow remaining entirely within the animals. During the turns, they are steady currents, as the path is resistive rather than capacitive, Kalmijn (1984). Whether the linear transient and circular steady motionally induced electric currents are sensed by animals in any way, we do not know. Yet, it certainly is interesting, perhaps even promising, to contemplate the issue, as long as the underlying physics and physiology are properly understood.

Jungerman and Rosenblum (1980) contemplated whether the vertical semicircular canals of the inner ear might detect the circular currents induced by interaction with the horizontal component of the earth's magnetic field when the animal turns about its vertical axis. Though a reasonable thought, physically worked out by them in great detail, the structural data available concerning the semicircular canals make it all but unrealistic, as I have explained in Kalmijn (1984). It would require the cupula of the ampulla to block the canal electrically so well that its resistance is high compared to that of the fluid core of the canal. However, there is no reason nor expectation for that to be true, as the cupula is purely gelatinous. I would have expected the authors to conduct the necessary measurements to prove their thinking correct before continuing to publish on it, as they did it in their subsequent Rosenblum et al. (1985) article. Recently, enticing new evidence has been brought forward by Nimph et al. (2019). However, the questionable physical validity of the idea has still not been properly addressed. Hence, let us leave the door to the truth ajar. Perhaps, there is more to it than I presently see. That would be exciting news.

The magnetic sense may detect the earth's magnetic field per se or merely its gradients

Since magnetic anomalies have been a cause of many misconceptions, let us start with their formal definition: magnetic anomalies are the vector or scalar differences between the measured *magnetic* field and an estimate of the main field. The anomalies are due to the magnetic properties of rocks in the earth's crust. The main field is caused by the fluid motions in the earth's core. Detection of the earth's magnetic field *per se* means that the animal perceives the local magnetic field as if it were perfectly uniform, represented by just one three-dimensional magnetic field vector. This is in contrast with animals detecting the nonuniformities, that is, the scalar or vector gradients in the local magnetic field. The two modes of magnetic sensing are complementary, providing different, highly valuable information. It is often suggested that animals may not rely on the magnetic field *per se* but only on the magnetic anomalies, that is, on the local deviations from the main field. However, since the main field is an inferred field, the anomalies can only be calculated, not directly measured as such. What can be measured are the nonuniformities in the local magnetic field in terms of gradients.

For the following discussion, we should be very much aware precisely what the animals detect, the magnetic fields *per se* or their nonuniformities, and what they may learn from the information they receive. The earth's magnetic field *per se* is given by the vector **B** at each point in space, that is, by its scalar magnitude *B*, the strength of the field, *and* its direction three-dimensionally. In regions where the magnetic field is conservative, the magnetic field may equivalently be given by a scalar magnetic potential φ_m at each point in space, where φ_m is defined by the line integral $\int \mathbf{B} \cdot \mathbf{ds}$ of **B** and, *vice versa*, **B** is defined by the spatial gradient $\nabla\varphi_m$ of φ_m. Although the magnetic strength *B* and the magnetic potential φ_m both are scalar quantities pertaining to the very same magnetic field **B**, they stand for very different features of the field. Hence, also the spatial gradients of the scalar quantities φ_m and *B* are very different vectorial quantities. Thus, $\nabla\varphi_m$ stands for the direction and magnitude of **B** in regions where the magnetic field is conservative, whereas ∇B stands for the direction in which the strength of the magnetic field changes the steepest and how steep that is, independent of the direction of the field. It is like a cyclist not caring which direction they go so with respect to the direction of the wind, but seeking the direction in which the strength of the wind increases or decreases with distance steepest, and how steep that is, independent of the direction in which the wind is blowing. To complete the list of pertinent definitions, the gradient of the vector **B** in regions where the magnetic field is not conservative is a second-order tensor giving the vector change in the vector in **B** when going from the position under consideration into a chosen direction, which is again is quite different from the gradient of *B*. Bewildering to say the least.

The strength of **B** measures on the order of 40,000 nT, whereas the strength of the anomalies measures on the order of 400 nT. A 3-D fluxgate magnetometer measures both the strength and the direction of a magnetic field, whereas a proton precession magnetometer measures only the strength of the field but does so at much greater sensitivity. Actually, what the anomaly maps show are mostly small differences between the large measured and inferred main fields. Hence, without knowing the main field, the anomalies have formally no absolute meaning. What the animal detects is the measured field or its scalar or vector derivatives, not the main field, nor the anomalies. One thing is clear: the ampullae of Lorenzini are perfectly well designed and more than sensitive enough to detect the earth's magnetic field **B** *per se*, but not at all designed and not at all sensitive enough to detect the local nonuniformities in the measured field. For that, the animal requires a magnetic sense on a par with a proton-processing magnetometer. Note that what people commonly refer to as anomalies, or rather as local nonuniformities, meaning the local spatial structure of the magnetic field, does not offer a compass sense, as it inherently does not indicate which directions are north, south, east, or west. Instead, they provide for a topographic map sense for animals familiar with the terrain.

A laudable field test pertinent to the concept of magnetic field gradient. Klimley (1993) made extremely impressive night-time hammerhead shark trackings in the Sea of Cortez, which I was so fortunate to witness in person. However, in his article he argues against the use of motional-electric fields as the guiding cue for the hammerhead sharks on their long journeys. Since he focused on a correlation between the sharks' travels and the local magnetic intensity gradient map, I agree, based on his hypothesis, the use of the ampullae is all but out of the question. Yet, even his magnetic intensity gradient hypothesis does not exclude the concurrent use of the animals' compass sense based on the full strength of the local earth's magnetic field. The results of Klimley's daring night-time trackings, made with his small panga using custom-made acoustic telemetry tags, will forever remain enigmatic. Unfortunately, NSF support for the continuation of his unique fieldwork, in collaboration with me to provide a more rigorous physical foundation as to the use of the sharks' electric sense, was not granted. This was a lost opportunity!

4.4 Kirschvink's Errors

Up to this stage, the differences of opinion between those of my peers and myself were within the reasonable range of misunderstandings and debate. However, from now on, the situation gets more critical. The attacks on the validity of my research are such that, if justified, would fatally undermine my research. Hence, I have no choice but to expose the errors of my critics. I want to give the next generation of biology and physics students the insight needed to correct these errors and expand on the field of research that I pioneered in deference to the sharks and rays, for which I have the greatest respect.

A first in-vain attempt to deny sharks and rays their motional-electric magnetic sense

Kirschvink (1989) was so eager to disqualify my motion-electric interpretation of the long series of magnetic orientation experiments I conducted in the stingray, *Urolophus halleri*, that he proposed that the animals may not have oriented to the earth's magnetic field *per se* but instead may have relied on the local non-uniformities in the magnetic field, even though the test fields were uniform within less than 1%. I considered such a precision quite good for a 5 m in diameter Helmholtz coil system enclosing the 1.5 m in diameter observation area in the shallow seawater observation tank, sized to make the ambient magnetic field large enough to give the stingrays sufficient space to determine the direction of the field and amply uniform to exclude the possibility that the animals detect the local non-uniformities in the field based on the principle of motional electric induction. Hence, in my opinion, the test fields applied were more than adequate for the very first decisive test to determine whether the animals are able to orient in the earth's magnetic field, a possibility I had logically inferred based on the animals' astounding electrical sensitivity and the physical principle of motional electric induction. Nevertheless, Kirschvink seized on the slight non-uniformities, by plotting, not the uniform fields as such, but their grossly amplified deviations from uniform, as if they were fatal flaws, and offered a convoluted alternative explanation for the highly significant results of the experiments, an explanation for which he did not present the slightest evidence. Yet, the experiments were conducted such as to make his convoluted scenario in advance extremely unlikely even if the animals would have—as they may—an *additional*, much more sensitive magnetic detection system to detect the local magnetic non-uniformities.

Certainly, less than 1% percent of magnetic non-uniformities would have been utterly undetectable based on the induction principle. Yet, Kirschvink proposes that (1) the rays may have ignored the motionally induced electrical signals they receive day and night when moving in their ocean habitat and in my land-based observation tank in which the experiments were conducted, to wit the motional electric signals offering them unambiguous sensory information about the ambient electric field as well as about their magnetic compass headings, (2) instead may have relied on another, unknown kind of magnetic sensory system, at least a hundred times more sensitive, capable of detecting the weak magnetic nonuniformities, (3) may have learned to associate the presence or absence of the nonuniformities (or in trials, in which the vertical component of the earth's magnetic field was nulled, the *difference* in nonuniformities) with some kind of visual, chemical, or tactile cues in the diffusely, dimly lit observation tank, to enter either the western or eastern feeding station to secure food. They would have had to accomplish this, even though odor cues worked against the training and between trials the bottom sand was thoroughly raked, and even though the magnetic field was after each trial extremely slowly reduced to zero and extremely slowly increased to natural strength again, randomly either normal or reversed, to prevent detectable electrical transients. Kirschvink admits, for sure, the animals oriented magnetically. However, without evidence he offers such a far-fetched alternative explanation of how the animals detected the magnetic field. Unfortunately, many subsequent scientists have uncritically picked up on Kirschvink's erroneous theory.

A matter of principle

The final proof for the existence of a sensory system rests upon conclusive behavioral evidence. In my opinion, merely investigating the changes in behavior after sensory impairment, such as ablation of the pertinent sense organs or attaching impediments like permanent magnets, are very dangerous methods. As a graduate student, I reluctantly agreed with Dijkgraaf together selectively to denervate the ampullae of Lorenzini to prove that they are the sense organs mediating the responses to electric fields. We may have done so convincingly, but I still do not feel proud of it. In all my later behavioral experiments, I have adhered to my common-sense principle that the electric and magnetic fields applied to test the animals' behavioral responses shall be either naturally occurring fields or faithful simulations thereof. In that vein, I elicited the electrically guided feeding responses of sharks and rays by presenting them with the bioelectric fields of live prey or multipole field simulations of similar strength and frequency content. I furthermore elicited the open-ocean orientational responses to electric and magnetic fields always to essentially uniform electric and magnetic fields of natural strengths. True, in my ambient-electric and earth-magnetic orientation experiments, I had to resort to conditioning procedures. Yet, I conditioned the animals again to naturally occurring electric and magnetic fields or faithful simulations thereof and by asking the animals to execute orientational behaviors that I may reasonably expect them also to perform in daily life. I question the validity and applicability of behavioral experiments using stimuli that do not occur in nature, either in configuration, strength, or in time course, even though the literature on the electric and magnetic sense is plentiful of them.

The Electromagnetic Research Facility at the Scripps Institution of Oceanography

The Electromagnetic Research Facility, EMRF, I designed and built in a remote area of the Scripps Institution of Oceanography campus features a 12 ft in diameter, 7 ft deep cylindrical seawater observation tank in which the ambient magnetic field can be made as it prevails in any part of the worlds' oceans with a uniformity within 5 nT. No surprise, Kirschvink considers the design of the new Electromagnetic Research Facility flawed as well, since I did not implement his 1992 proposed bifilar and electrostatic shielding embellishments. I easily could have done so, but the coil system I devised operates perfectly well without, as I recently showed him.

A historical note

At the time of Kirschvink's (1989) vindictive article, there were two cases of magnetic orientation known in which not only the orientation capabilities of the organisms were well-established, but also the proposed detection mechanisms were physically sound and clear. The first one concerns the magnetic sense of sharks and rays, theoretically predicted by Kalmijn (1974) based on the physics of motionally induced electric fields and the high electrical sensitivity of the animals, and behaviorally confirmed by Kalmijn (1978b, 1981) in a protracted series of conditioning experiments and by several incidental observations. The second case concerns the magnetic bacteria, fortuitously observed by Dr. Richard Blakemore (1975), at the time a younger microbiology colleague of Kalmijn at the Woods Hole Oceanographic Institution, Massachusetts. In a collaborative effort, Kalmijn and Blakemore (1978) and Kalmijn (1981) experimentally studied the forced movements of the bacteria in magnetic fields and determined the dipole moment of their magnetosomes in experiments using magnetic pulses to reverse their polarity. It was then that Kirschvink (1989), Gould (1998), and many other scientists entered the fray, many of them adding exciting new data but also a lot of noise and dust.

4.5 Did Faraday, Maxwell, Einstein, and the Creator Get It All Wrong?

The new erroneous physics and biology of motional electric orientation

A subsequent article, published by Paulin (1995), eclipsed Kirschvink's critique. It was meant thoroughly to destroy my "nonsensical" magnetic orientation theory and to replace it with a "new, less esoteric" one. Paulin mercilessly attacks my 1984 "Further Analysis," but refers to my original 1974 dissertation and its subsequent 1981 sequel only in passing and not at all to my ONR 1978 and relativistic 1988a articles on the topic, even though in them he could have found ample evidence preempting essentially all his allegations. I have taken Paulin's article very seriously to rescue those who fell victim to his physical and physiological aberrations. In a separate article, I have written a full-length reasoned rebuttal. Here, I will merely highlight a selection of his most blatant misconceptions, resulting from ignoring well-established facts and violating sound scientific reasoning.

Why reading the original literature is so important

Paulin berates me for speaking of sharks and rays detecting DC electric fields, even though their electroreceptors, the ampullae of Lorenzini, are not DC, but low-frequency AC receptors, *as if that never occurred to me*. He offers this to be the reason for writing his article, thereby exhibiting his ignorance of the pertinent literature. In response, I will give three excerpts, available to Paulin already long before he wrote his virulent article. True, my earlier statements are old—so am I—but certainly not outdated. They were not presented in hindsight, but with confident foresight!

1. Kalmijn (1974), p. 185:

> That sharks and rays are most sensitive to frequencies below 8 Hz is in good agreement with Murray's (1962, 1965) nerve recordings and Waltman's (1966) biophysical measurements. However, electrophysiologically, the ampullae of Lorenzini do not behave as true dc receptors, since they adapt almost completely within seconds. This seems to contradict the fact that feeding responses are evoked most readily just with dc fields. It actually means that, to detect dc dipole fields, sharks and rays must move with respect to them in order to bring the eventual stimuli within the frequency range of the electroreceptors.

2. Kalmijn (1974), p. 186:

> For orientation with the aid of inanimate electric fields, modulation of the stimulus is again necessary. To perceive the uniform dc fields occurring in the ocean, the fish could take any

> non-translational course of swimming. Their natural tendency to swerve right and left could conceivably suffice. For the detection of their own motional electric fields, translational accelerations may do as well.

3. Kalmijn (1988b), p. 176:

> But why the low-frequency cut-off? It is the dc-instability in the time domain or the low-frequency noise in the frequency domain that makes dc measurements unwise to attempt unless absolutely necessary. One solution is to shift the near-dc stimulus into a higher, less noisy, frequency band, for which the normal movements of the recipient animals may again suffice. Thus, the fishes' swaying motions could be considered either as the carrier for the stimulus or as the sampling function needed to realize the desired frequency shift. In predation, either the prey or the predator has to move. Without motion can neither be detected nor caught.

True, the ampullae of Lorenzini are not DC *sensors*, but sharks and rays nevertheless prove behaviorally to have a DC *electric sense*, day and night detecting the DC electric fields of prey and ocean currents. All *sense organs* monitoring the world around living creatures are part of *sensory systems* that include the exploratory movements made by the organism or parts thereof. It is the behavioral feedback, under control of the pertinent brain centers, that enables the peripheral sensors to perform their detection tasks. It is not difficult to think of examples of this. Even our eyes would not allow us to see well if they were not micro-jittering continuously. At a larger scale, our inner-ear inertial sense organs are not stimulated unless we move or are moved. Interestingly, to detect the earth's magnetic field, sharks must not only move to bring the DC stimuli into the low-frequency range of their electroreceptors but also to produce the very stimulus itself to begin with. As I explained in 1974, in detecting the electric fields of ocean currents, sharks operate their electric sense in the passive mode, whereas in detecting the electric fields they themselves motionally induce, sharks operate their electric sense in the active mode. In both modes, the animals must perform the proper exploratory movements to detect and interpret the electrical stimuli.

Which physics?

Paulin (1995) sets the stage for his critique on my theory of electric and magnetic orientation in sharks and rays by first outlining his own misunderstanding of the physics involved based on the undergraduate physics textbook by Serway (1990) and, in particular, on the introductory course in electromagnetism by Hammond (1985). Even to those, Paulin does not strictly adhere. With due respect, these textbooks are not at all at the high, but demanding level of, for instance, Panofsky and Phillips (1962, 1972). Biologists shall be aware that sharks and rays practice intricate physics and can teach us scientists a lesson or two. Yet, their intricate physics has so far not explicitly entered the textbooks. If I say that Einstein (1905) was fully aware of them, people wonder.

Which ampullae of Lorenzini?

Paulin's model of the ampullary system considers the skin pores and ampullary canals to funnel the motionally induced electric current through the sensory epithelia, whereas—to the contrary—the induced electric current flows across the animal through the skin and body tissues, bypassing the pores and ampullary canals, as the sensory epithelia lining the blind ampulla proper are ideal voltage-sensitive devices, not *drawing* any induced current. Actually, the resistance of the sensory epithelium is slightly negative, causing a slight stimulus-enhancing current to flow in the opposite direction. Hence, Paulin's ampullary model is structurally and physically incorrect. This is not a matter of opinion but a well-established fact, first proven beyond doubt by Waltman in 1966. Hence, in contrast to Paulin's model of the ampullary canal acting as a moving rod driving an induced electric current through the stationary seawater part of the circuit, instead it is the skin and body tissues that act as the moving rod driving the induced electric current through the stationary seawater part of the circuit, whereas the ampullary sense organs measure the resulting voltage between skin pores on opposite sides of the animal without themselves passing any induced current.

Moreover, in detecting the oceans' ambient electric fields and in detecting the earth's magnetic field, the ampullae of Lorenzini would, according to Paulin, physically operate entirely differently, for which there is no evidence. If you ask the animal, the electric fields it receives from drifting in ocean streams and from inducing itself when traveling appear both as imposed uniform electric fields, differing only in how they respond to the animals' exploratory movements. To follow Paulin's reasoning, one must keep in mind his incorrect physical ampullary model on which his arguments are based. No wonder, he runs into many more discrepancies. If he had more diligently scrutinized the pertinent literature, he could have realized his error and spared us this unnecessary controversy.

Forget about the charges, they have no place in Paulin's model

Founded on his faulty "more precise" model, Paulin arrives at the conclusion that sharks, in orienting to the earth's magnetic field, detect the $\mathbf{v} \times \mathbf{B}$ in the ampullary canals, which is fundamentally incorrect, as I showed in the very first 1974 introduction of my motional electric orientation theory. True, in my lighter version of the theory, meant for the layman reader, I considered the $\mathbf{v} \times \mathbf{B}$ to provide a reasonable estimate of the physically adequate stimulus $-\rho\mathbf{J}$ sharks detect, with the warning that the $\mathbf{v} \times \mathbf{B}$ is fundamentally different from the $-\rho\mathbf{J}$, both in its physical nature and in its size. Though approximations can be helpful, they can also be dangerous, as they have limited validity: here only to sharks in ocean waters, not to sharks entering brackish or fresh waters. It applies even less to freshwater animals, not to speak of birds flying in thin air. Moreover, in the animal's frame of reference, there is no $\mathbf{v} \times \mathbf{B}$ in the ampullary canals, since it is the environment that moves, not the animal, rendering the issue moot. When introducing the lighter version, I made its limitations manifestly clear. Even so, though the lighter version was meant to be conceptionally helpful, it caused Paulin to ignore the fundamental charge accumulations on opposite skin surfaces of the animal, the electric field of which not only drives the return current through the surrounding medium but also detracts from the $\mathbf{v} \times \mathbf{B}$ inside the animal's body and ampullary sense organs, leaving the $-\rho\mathbf{J}$ for the animal to detect. True, in marine sharks and rays the charges diminish the effective stimulus only little, but in freshwater, terrestrial, and airborne animals a lot. Hence, I repeat for emphasis, the lighter pared-down version is only meant to give the layman biologist a less demanding, informal feel for the theory of electric and magnetic orientation in sharks and rays, not as a starting point for the academic professional who chooses to analyze the theory in physical detail.

A Babylonian confusion of tongues

Warning! Paulin invents his own terminology—instead of wisely adhering to that of the physical oceanographic literature, his physics textbooks, or my writings—which make his reasoning very difficult to follow. He speaks of *the force* **F**, where Longuet-Higgins speaks of the *motionally induced emf per unit distance*, and I, equivalently, of the *motionally induced electric field* $\mathbf{v} \times \mathbf{B}$, while Hammond, Paulin's textbook, speaks of the *motional electric field,* commonly reserved for the electrostatic field of the charge accumulations resulting from the induced electric current. Paulin also speaks of the *induced electric field*, where Longuet-Higgins and I speak of the *motional electric field,* $\mathbf{E} = -\nabla\varphi$, that is, the field of the induced charge accumulations, which Hammond, Paulin's textbook, calls the *electrostatic field.* This all has led to a Babylonian confusion of tongues. True, the term *induced electric field* is commonly used as a shorter version of the *motionally induced electric field* $\mathbf{v} \times \mathbf{B}$, and the terms *motionally induced* and *motional electric* are often used generically, without specifically meaning either specific kind. Hence, the meaning of his scientific terms must become clear from the context. You just have to figure out yourself what Paulin means by what he says.

Sharks know better

Paulin argues incorrectly that, since it is difficult for sharks to interpret the oceans' ambient electric fields, it is nearly impossible to correct for them in order to establish their magnetic compass headings motion-electrically. Here, Paulin attacks a fictitious model of a shark of his own making, since the motional electric detection of the earth's magnetic field does not at all depend on the interpretation of the ambient electric fields. From the very 1974 beginning, I have emphasized that the electrical signals received from environmental electric fields and the electrical signals received from the shark's own motion in the earth's magnetic field are readily kept apart, as the two kinds of fields, though similar, are fundamentally different functions of the animal's exploratory movements, as I have said so often. Rather than to compensate for the ambient electric fields, the animal uses the ambient electric fields to orient in ocean streams and adjacent waters, while at the same time using the fields they themselves induce to sense their magnetic compass headings. Paulin argues against the use of the ambient electric fields based on my statement that "interpretation of the various fields requires considerable expertise on the part of the animals," as if not all sensory systems would require considerable expertise on the part of animal and men, for which they have brains.

Sharks live in the shark world, not ours

Another fundamental issue of general interest is Paulin's confusion as to the reference frame in which the ampullae of Lorenzini operate. An observer at rest in the earth frame may consider the ampullae to move and thus motionally to interact with the earth's magnetic field, thereby inducing an electric field. Yet, sense organs collect the information concerning the outside world as presented in the frame of reference in which the organs are at rest. Hence, a shark receives in its own frame of reference, only electric fields and has no knowledge of—what we call—a magnetic field, unless it has an additional, still unknown, non-induction magnetic sense. The only difference between the electric fields the animal receives from ocean currents and from the fields itself induces, I repeat again, is their dependence on the exploratory acceleration and turning movements made by the animal. If the ampullae are optimized for the detection of the one kind of field, then they are optimized for the other kind as well. To speak of electric and magnetic is merely the human convention of describing different physical features of one and the same electromagnetic fields.

But does not the animal orient in the earth frame of reference? Yes and no. To get around in the earth frame of reference, it has to do the detection, data processing, and orienting in its own frame of reference and subsequently transform the information to *infer* its behavior in the earth frame of reference. Whether sharks really do so, I doubt it, even though unwittingly we do so all the time! Paulin also denies Einstein's special theory of relativity to be fundamental to the electromagnetic orientation of sharks, even though the railroad train equivalent of the shark's electromagnetic solution of the problem was for Einstein in 1905 the reason to develop the special theory of relativity. We feel very rueful for the sharks and for Einstein to be so poorly understood. Do not underestimate my scientific heroes.

Paulin throws the significant signal out to study the left-over crumbs

When it comes to identifying the physically relevant features of the stimuli fundamental to the shark's magnetic compass sense, Paulin single-mindedly focuses on the swerving movements of traveling sharks that I have repeatedly referred to in the past as a confluent complement of the exploratory movements needed to detect the ambient electric field as well as the earth's magnetic field. However, Paulin disposes of the strongest signal, the DC content resulting from the animal's forward velocity, since, after traveling for a while, it is not transduced, at least at a precise constant forward speed, wasting precious information during the initial period of acceleration. As for the remaining, to a certain extent distorted sinusoidal signals the animal receives, he does not rely on them as such but only how their second-order wiggles and ripples deviate from those of the signals concurrently received from the semicircular canals, while ignoring the sizable amplitudes of the first-order quasi-sinusoidal signals providing the animals with all electrical information needed to determine their magnetic compass headings. To make the time-course comparison, he first normalizes the motionally induced signals in order to make the deviations from sinusoidal of a similar size as those of the responses from the inner-ear semicircular canals, hence *discarding the most prominent and informative features of the signals, to wit, not only the DC content, but also the directionally sensitive amplitudes of the fundamental sinusoidal contents in favor of the left-over crumbs.*

In short, when traveling quietly, swerving through small angles, sharks would—according to Paulin—only use a tiny subset of the electrical stimuli available. Unfortunately, the scaled left-over fine structures of the already not very large signals are so small that it is all but inconceivable for the central nervous system to recover them from below the noise level. Why would the animal ignore the tell-tale signals to find the solution in the left-over crumbs? Moreover, in detail, the leftovers appear to be more the result of the simplifying assumptions made concerning the forward motion and sinusoidal swaying of the head than that they are characteristic of the animals' true mode of travel. They are even smaller than the interfering few percent variations in the forward speed of the animal as a whole that I measured from video recordings of the animals' steady swimming. They are also quite different from those of an equivalent model I introduced, assuming the forward speed along the path of travel to be constant, or to fluctuate in a known manner, rather than the forward speed in the direction the head moves to be constant. Paulin's model is based on contrasting assumptions, whereas my alternative model is logically correct, though needs real-live video data to verify it quantitatively.

Back to 1974

Originally, in 1974, I based my theory of motional electric orientation in sharks and rays on the characteristic seemingly haphazard accelerating and turning motions the animals commonly make, enabling them to detect the electrical stimuli from interaction with the ambient electric and earth's magnetic fields in their full strength and to do so on short notice. I mentioned also the possibility of sharks detecting their magnetic compass heading based on their characteristic swerving mode of traveling, but why did I not analyze the latter course of action in the first place? The answer is logical and simple. I considered the swerving mode of swimming to consist of a concatenation of smoothly overlapping individual turns to the left, turns to the right, linear accelerations, and linear decelerations, each individually giving the animals full information about the ambient electric and magnetic fields, respectively. Hence, to analyze the concatenation of the individual moves, it made sense first to analyze the individual moves in isolation, with the expectation that the animals follow that logical course of action. Obviously, the smooth concatenation of individual moves is a compromise between effective travel per se and sensing of magnetic heading. During regular travel, the turning and acceleration motions are certainly not optimal for sensing magnetic headings.

Imagine, a shark decides to travel in a certain magnetic direction. How shall it proceed? Well, *first it shall determine which the desired direction is, then start swimming and maintain that direction.* How shall the animal determine the desired direction? Obviously, by making the most efficient individual acceleration and deceleration moves, such as those I have described. How shall the animal then maintain that direction? Easy, by keeping all the electrical information it receives from both the ambient electric and from the earth's magnetic fields constant, however complicated the total electrical input may be. But if there is any uncertainty whether the circumstances during travel have changed? Just make again a few individual exploratory movements, perhaps now and then take a deep dive to explore the electrical depth profile, and then resume swimming in the desired direction again. In short, during the steady travel, there is no need to determine the compass heading, there is only a need for maintaining the intended compass heading, which is a much less demanding task.

In short, yes I believe that sharks during regular travel use the motional electric signals they receive from their swerving motions to be guided by the ambient electric *as well as* in the earth's magnetic field, but not by recovering Paulin's model-particular leftover signals from the noise of the sensory system, but by focusing on the more prominent features, not to discover the compass heading but merely to maintain the compass heading, in the most convenient way. Sure, the otolith organs and the semicircular canals of the inner ear serve, by means of feedback, to control the linear and angular accelerations of the animals' exploratory movements, but for the animals to rely on the difference in time course between the inner-ear signals and those of the ampullary sense organs seems unnecessary and, since the fine structures of the signals are hardly above to well below the noise level of the system, whereas the amplitudes of the sizeable sinusoidal fundamentals, which Paulin ignores, offer all the information the animals need. *Might not animals in general make a procedural difference between determining their compass heading and maintaining that compass heading, although the two procedures may well overlap?* That would make orientation on long-distance travels a much less complicated endeavor.

The difference between orientation and navigation
As a last allegation, Paulin faults me for failing to explain the sharks' electric and magnetic *navigation* abilities, an issue I never addressed. How could I have failed in an effort I did not engage in? I have only shown that sharks (1) receive and very well may use the electric fields they motionally induce to sense their magnetic compass headings and (2) receive and very well may use the electric fields motionally induced by ocean streams to detect their drift, both for the purpose of open-ocean *orientation*. Whether and how sharks apply their electric and magnetic orientational abilities to the difficult task of navigating the worldwide oceans, I will leave it to the imagination of the educated reader. Yet, without knowing their orientational abilities or, even worse, having a wrong notion of their orientation abilities, it will be hard to find out how sharks travel open-ocean waters. Navigation studies shall take all sensory modalities and brain algorithms into account, more than I can handle. Nevertheless, evolution made it perfectly sensible.

4.6 Stick to the Proven Physical Facts and Principles

A second in vain attempt to deny sharks and rays electrically sense their magnetic heading
Kirschvink et al. (2001) desperately attempted to malign the motional electric principle of magnetic orientation a second time. Apparently inspired by the magnetic bacteria he got in 1978 from my Woods Hole Eel Pond stock, Kirschvink seems to entertain the idea that all animals *might* orient magnetically by means of magnetite, including sharks and rays. Of course, sharks and rays *may* have an *additional* magnetic detection mechanism capable of sensing local magnetic non-uniformities, but

my 1978 experiments do not give any indication whether such an alternative would be involved. I say, an "*additional* magnetic detection mechanism," as it would be nonsensical for the animal with great effort to ignore the unambiguous and readily detectable signals they receive motional-electrically to give priority to an unknown magnetic non-uniformity detector. Even more bizarre, Kirschvink, Walker, and Diebel (2001) consider *induction in the earth's magnetic field unwanted noise for the electric sense,* and infer that sharks and rays *should* have a separate magnetic sense, conceivably based on magnetite, to suppress the electric signals they receive by moving in the earth's magnetic field, to prevent them from interfering with the animals' zeroing in on the bioelectric fields of their prey. Kirschvink, Walker, and Diebel incorrectly insist that "the shark's electric sense is primarily used for locating prey." The authors and their followers apparently have a deeply ingrained aversion to the idea of induction-based orientation in sharks and rays.

Kirschvink et al. (2001) claim that *electrical induction as a mechanism for magnetic compass orientation in elasmobranch fish would be inefficient in navigating sharks without large structures*, many times larger than the ampullae of Lorenzini. Here, they call on Klimley (1993) who correctly argued that the ampullae of Lorenzini are not sensitive enough to detect the magnetic *nonuniformities* that the sharks he tracked in Mexico waters oriented to. However, my original theory of motional electric orientation concerned orientation with respect to the magnetic field *per se*, not with respect to the orders of magnitude weaker magnetic gradients. In the first instance, old-time mariners used to navigate relying on magnetic compasses indicating the direction of the earth's magnetic field *per se*, as Kalmijn's stingrays evidently did in his behavioral experiments. For that, the shark's ampullae of Lorenzini are amply large and sensitive enough for the animals to sense their magnetic compass headings. True, they clearly did not evolve to sense motion-electrically the magnetic nonuniformities other animals *may* rely on. Let us leave that to Kirschvink *cs*.

Then, Kirschvink et al. (2001) claim to have experimentally proven that stingrays do not orient in the earth's magnetic field motional electrically, only to exhibit their lack of understanding concerning the physics of motional electric induction. Actually, the results of their experiments are fully consistent with the motional electric principle on which Kalmijn's theory is founded, whereas their interpretation of the "falsifying" experiment that Kirschvink et al. (2001) conducted might cause Faraday to turn over in his grave. Ironically, their experiments unwittingly presented excellent support for my views despite their intent to the contrary.

Indeed, the lack of a thorough understanding of the physics underlying the biological role of motional electric stimuli is patently evident from the statement made by Kirschvink et al. (2001) that *"an attached magnet, moving with an animal, should not impair an induction-based sensory system, whereas it should disable any other receptor within its field of influence."* Two years later, Walker et al. (2003) explain: "*Because attached magnets impose a constant field relative to the body, they will make no contribution to the electrical signal induced by an elasmobranch fish as it swims. Magnets should therefore not affect magnetic detection using electroreceptors, but they should impair magnetite-based magnetoreception provided they are placed close enough to the magnetite to interfere with its use in magnetic field detection.*" Since the magnetic orientation of their stingrays was lost after they attached the magnet, Walker, Diebel, and Kirschvink concluded that before the magnet was attached the animals must have relied on another sensory system than induction-based magnetic sensory system.

Kirschvink, Walker, and Diebel apparently were not aware of the offending motionally induced electric currents, not produced by the motion of the magnet with respect to the sense organs, but by the motion of the magnet relative to the seawater, with respect to which the animal moved. These fields will crudely act on the animal and directly interfere with the signals received from moving in the earth's magnetic field. Imagine the magnet to be oriented along the ray's left-right, transverse body axis. The magnetic field directly above and below the animal will then be directed horizontally. By interaction with the water with respect to the ray and the magnet move, a vertical electric current will be generated, stimulating the animal along its ventro-dorsal axis, the axis along which it is supposed to detect the electric field resulting from its forward motion in the presence of the horizontal component of the earth's magnetic field. In short, they made the same old mistake of Lissmann and Machin (1958), to be perpetuated by Molteno and Kennedy (2009), to consider the local $\mathbf{v} \times \mathbf{B}$ at the position of the sense organs to be the physically adequate stimulus, rather than the local $-\rho\mathbf{J}$ resulting from the motionally induced $\mathbf{v} \times \mathbf{B}$ and the resulting electric field $\mathbf{E}$ throughout the whole water volume with respect to which the animal moves. Yet, in their concerted effort to disprove the animals' reliance on the motional electric fields they receive, the results of their experiments are in full support of just that what Kirschvink, Walker, and Diebel tried so hard to disprove.

Ironically, Walker et al. (2003) seem to rely more on their close colleagues who unwittingly misunderstood Kalmijn's writings than on the original writings of Kalmijn himself. A few relevant examples shall be called into question here. Relying on Bodznick et al. (2003), the authors believe that in general the ampullae of Lorenzini differentially detect the voltage differences between the skin pores, which seems to contradict the fact that sharks—in orienting to the earth's magnetic field—receive the strongest motion-electric signals when the voltage difference between the skin pores is closest to zero. Did not Kalmijn (1974) explain that the ampullae of Lorenzini detect the pore-to-pore line integral of $-\rho\mathbf{J}$ due to the induced electric current across the body in orienting to the earth's magnetic field Walker et al. (2003) were talking about, which indeed is the larger the more the shark is electrically shorted by the seawater medium.

To make matters even worse, Walker et al. (2003) refer to Paulin (1995) to blow new life in my original theory, saying "*...further theoretical analysis was needed to show that the low-frequency movements of the body during swimming will still permit extraction of the information about magnetic field detection.*" Hence, Walker, Diebel, and Kirschvink are seeking solace from Paulin's error-ridden article, as did Montgomery and Bodznick (1999), Bodznick and Montgomery (2005), who I had expected to know better. However, Paulin did not solve any problems, he only made them, violating established physics and biology, disregarding the well-known operation of the sense organs, and depriving the animals of the rich information they receive in favor of the distortions introduced by his model. Understandably, Walker, Diebel, and Kirschvink (2003) got thoroughly confused after reading Paulin's article. Actually, Paulin merely plagiarized my reasoned inference that even quietly traveling sharks may motional-electrically sense their magnetic compass headings, calling it *his new theory*, but then completely erred in the physical basis of his own version. Walker, Diebel, and Kirschvink are very consistent in perpetuating their errors. They still maintain in 2007 that Paulin *rescued* Kalmijn's theory by stating, I repeat, *the low-frequency movements of the body during swimming will still permit extraction of the information about magnetic field detection*," exactly what Kalmijn originally said—true long ago—which apparently escaped Walker, Diebel, and Kirschvink altogether, as if I have not often enough made it clear over the years:

> 1981, p. 1115: "While swimming, sharks normally swerve alternately right and left, thereby scanning the induced field over a certain arc."
>
> 1984, p. 547: "The swimming movements modulate the stimulus fields, bringing them within the frequency range of the receptors."
>
> 1988a: "The fishes' swaying motions could be considered either as the carrier for the stimulus or as the sampling function needed to realize the desired frequency shift."

Dear reader, ask the sharks their opinion in proper behavioral experiments or consult with our trusted teachers Faraday, Maxwell, and Einstein. Then, decide for yourself.

4.7 Two Review Articles Misrepresenting the Original Literature

Why do so many ill-conceived ideas in science linger for so long in the literature? The ideas that appeal to the less learned readers tend to be agreed upon, whereas those that do not speak to their imagination tend to be rejected, especially when it concerns review articles based upon hear-say rather than unbiased inquiry. But why would the reader shy away from assessing the original ideas on their intrinsic merits? That is exactly the issue. Most often, in particular in the instance of my electric and magnetic orientation theory, the reader may not have the physical and biological background to fully understand the intricacies. We cannot blame him or her for that, as the motional electric physics is rather exacting. We can only blame the author of a correct theory not to have made his ideas clearer and blame the author of a false theory to have defended it too glibly. My rebuttal is intended to separate the hard science from the make-believe stories, not by making the presentation more eloquent and authoritative, but by rendering it more comprehensive and reasoned in intelligible scientific language. It may on short term defeat its purpose by asking too much from the reader. Yet, in the end, the truth will emerge and universally prevail. I sincerely hope that my article will steer the reader in the right direction. But do not take my word as indisputable, decide for yourself.

Importance of review articles to keep the layman up-to-date, but of what, sense or nonsense?
Scientists and science writers at times need information concerning the electric sense of sharks and rays, for instance in writing biology texts. They understandably resort to recent review articles to inform themselves of the state of research, not taking the time or having the expertise to consult the original literature. Hence, reviewers, especially the distinguished ones, have the holy responsibility to represent the literature correctly. In that vein, I feel obliged to comment on the review articles of Lohmann and Johnsen (2000), well known for their laudable studies on the magnetic orientation of sea turtles, and of Peters et al. (2007), who fell scientifically disappointingly short after they inherited my original Dutch electrophysiology laboratory when I left for the New World. The more the above authors present themselves or are regarded as the leaders in my field of research, the more their misunderstandings penetrate and get ingrained into the scientific literature. On the bright side, the beneficial effect of these perturbing review articles is that they oblige me, with trepidation, to identify and correct their misrepresentations. I wish I would not have had to do so, but so far nobody else has come forward, to my surprise and dismay.

4.7.1 First Troublesome Review Article

Why sharks and rays, but not turtles?
In their abstract and introductory paragraph, Lohmann and Johnsen (2000) incorrectly categorically downplay the ability of sharks and rays to orient in the earth's magnetic field by the principle of motional electric induction, saying "definite evidence that they actually do so has not yet been obtained" and "little direct neurobiological or anatomical evidence exists to support any of the proposed mechanisms." What they should have added is, "except that it is all but inescapable that sharks and rays orient motional electrically in the earth's magnetic field and that the anatomically and neurobiologically well-studied ampullae of Lorenzini are the pertinent sense organs." Sure, Lohmann & Johnsen qualify their statements, stating that "*primary* magnetoreceptors have not yet been identified unambiguously in any animal," as if *non-primary* magnetoreceptors are not magnetoreceptors, whatever *primary* means, see paragraph below under Sect. 4.8, "*Per se, direct, and motional electric magnetic detection.*" Even though my criticism will be relentless, I thank Lohmann & Johnsen for letting us know their misapprehensions, forcing me to correct them to the benefit of their peers and the scientific readership at large in the field of motional electric orientation. Note: I will only speak about what Lohmann & Johnsen say about my research, not about their highly respected research. Neither

will I speak of their comments on the research of other scientists in the field of magnetic detection in general.

Let me explain why their dismissive attitude is unwarranted and rests apparently more on their lack of understanding than on the sharks' questionable sensory abilities. No doubt, sharks and rays have the proven ability to orient in uniform magnetic fields of earth strength. No doubt, physically the animals motionally induce electric fields offering them complete information about their magnetic compass headings. No doubt, the ampullae of Lorenzini detect and pass the information on to the central nervous system. No doubt, physically the electrical stimuli the animals receive from moving in the earth's *magnetic* field are, in the animals' frames of reference, essentially of the same uniform configuration and of similar strength as the electrical stimuli they receive from the oceans' ambient *electric* fields. Hence, since their orientation in ambient electric fields is a well-established fact, their orientation in the earth's magnetic field based on the electric fields they themselves induce is all but certain, unless miraculously proven otherwise. Lohmann & Johnsen furthermore missed the crucial point that the detection of ambient electric fields and the electric fields the animals themselves induce by moving in the earth's magnetic field are physically entirely different functions of their exploratory movements and, thus, are readily detected side-by-side. Hence, I wonder if Lohmann & Johnsen have heedfully read and understood my several original writings on the topic. Or shall we believe that the motionally induced electric fields, although containing all the information needed for the animals to determine their drift in ocean streams and to sense their magnetic compass headings, are just a nuisance, as they would interfere with the electrical detection of prey?

Animals practice tough physics

Later in their article, it becomes evident why Lohmann & Johnsen have reservations about the induction procedure. Even though they dutifully credit me for introducing the proper physics, they nevertheless describe the detection of motionally induced electric fields in elasmobranch fishes as incorrectly outlined by Paulin (1995), apparently not aware that his alternative physics is not only at variance with the governing physical and biological principles I abide by in my electric and magnetic orientation theory, but also in violation of all respectable physics textbooks and the scientific works of our great teachers of the past, Faraday, Maxwell, and Einstein. Most blatantly, the gel-filled canals of the shark's ampullae of Lorenzini are not equivalent to conducting bars in which electric currents are induced when moving in a magnetic field, as Paulin (1995) claims, contrary to direct measurement by Waltman (1966) and Kalmijn (2003). The induced electric currents actually flow across the moving body, bypassing the ampullary sense organs altogether. Hence, not the ampullary canals but the animal's body is the equivalent of the moving bar, a fact Paulin vehemently denies. The ampullae instead act as high-ohmic voltmeters detecting the potential difference across the animal caused by the currents motionally induced in the skin and body tissues, not by the purported but nonexistent currents induced in the ampullary canals. The process of evolution has adhered to the physics of the creator, not Paulin's. Why did Lohmann & Johnsen not set the record straight by informing the reader of the fundamental disparity of the venerable old and erroneous new theory?

Resorting to a broken crutch, rather learning to walk themselves
Like Paulin, Lohmann & Johnsen wonder how sharks can tell which contribution to the total electric field they receive is attributable to their own motion and which is due to the motion of the surrounding seawater medium. As I have explained above, that is a moot issue, which Paulin did not solve but botched thoroughly. He tells us how sharks may ignore "the jamming effect [*sic*] of electric fields induced by ocean flows" to make it possible to detect their magnetic compass heading, rather than appreciating how traveling sharks may detect the ambient electric and the earth's magnetic field side by side, thus receiving complete electromagnetic information. He proposes that quietly traveling sharks (1) eliminate the DC component of the stimulus by adaptation of the sense organs, (2) disregard the amplitudes of the most informative fundamental AC signal they receive, (3) use only the polarity of the fundamental AC signal, (4) in the end resorting to the time course of the left-over distortions of the fundamental in comparison with the time course of the signals received from the semicircular canals, and that at levels hardly above to well below the noise level of the ampullary system. He furthermore ignores the need for sharks to correct for the electrical signals they motionally induce during their fast, electrically guided approaches to their prey in the presence of the earth's magnetic field. Moreover, from Paulin (1995), Lohmann & Johnsen incorrectly accept that "in induction models, difficulties arise because the animal would need to determine with great precision both its own velocity and the magnitude of the background electrical fields in its environment," which is merely an objection based on ignorance of the original literature.

Whether and how sharks and rays use their ability to orient in electric and magnetic fields for the purpose of *navigation* is another question. Yet, I can state for sure that sharks and rays crossing Bimini Bay kept a constant heading with respect to the ambient electric field Kalmijn (1988b). Moreover, I can say for sure that the stingray *Urolophus halleri* can decide to enter a feeding station in the magnetic west of their habitat and not in the east, or *vice versa* Kalmijn (1984). Let our lack of knowledge about their use of their electric and magnetic sense in *navigation* not distract us from the animals' phenomenal, well-established electric and magnetic orientational performances. Although navigation is not part of my electric and magnetic orientation theory, the animals' *orientation* abilities certainly are fundamental to their navigational skills. Without orientation, there can be no navigation.

Seeking dirt that boomerangs
Surprisingly, Lohmann & Johnsen take the baseless criticisms of Kirschvink *cum suis* of my magnetic orientation experiments seriously. To perpetuate Kirschvink's appalling remarks is, in my opinion, an equally serious insult to science as Kirschvink's remarks themselves. In reviewing the literature, scientists should in first instance consult the original sources and give explicit reasons as to why they consider one article more accurate than another. Next, they may want to get curious what other scientists think about the original sources.

Again and again, no relief in sight

Five years later, Johnsen and Lohmann (2005) dig their heels even deeper concerning the shark's motional electric detection of the earth's magnetic field, disavowing the all but self-evident. Let me quote Johnsen & Lohmann, though it is actually Paulin who speaks, all three arguments in utter contrast with my long-time teachings. Johnsen & Lohmann merely fall in Paulin's trap. Yet, my models are not that "simple" as to overlook several factors of fundamental importance, neither physically nor biologically.

> However, several factors significantly complicate these simple models. First, the electroreceptors of elasmobranches cannot detect the steady fields that were originally thought to arise. Second, the water surrounding marine fish is seldom motionless under natural conditions. Third, ocean currents are also conductors moving through the Earth's magnetic field, and so create electric fields of their own.

All the above "complications" were foreseen and dealt with in my very first 1974 presentation of the theory. Did Johnsen & Lohmann ever read my 1974 Springer Handbook chapter, page 185, which I wrote long ago but certainly is not outdated, to the contrary.

> Electrophysiologically, the ampullae of Lorenzini do not behave as true dc receptors, since they adapt almost completely within seconds. This seems to contradict the fact that feeding responses are evoked most readily just with dc fields. It actually means that, to detect dc dipole fields, sharks and rays must move with respect to them in order to bring the eventual stimuli within the frequency range of the electroreceptors. For orientation with the aid of inanimate electric fields, modulation of the stimulus is again necessary. To perceive the uniform dc fields occurring in the ocean, the fish could take any non-translational course of swimming. Their natural tendency to swerve right and left could conceivably suffice. For the detection of their own motional electric fields, translational accelerations and decelerations may do as well.

That is what at the time I proposed, have behaviorally proven to be correct, and ever since have explained over and over. Nevertheless, Johnsen & Lohmann, continue, or rather Paulin, saying:

> In principle, these problems might be overcome if sharks derive magnetic field information from the oscillating electric fields that arise as the ampullae on their heads move back and forth during swimming. Such fields should be distinguishable from fields caused by oceanic currents.

Ironically, the shark's swerving mode of swimming is a very effective way of orienting in ambient *electric* fields. Considering ambient electric stimuli *irrelevant* and ignoring them means discarding precious information. To the contrary, their swerving mode of swimming is actually a less efficient routine for a traveling shark to detect its *magnetic* compass heading. Any sizeable accelerations or decelerations and any sizeable changes in heading are more informative than just the short episodes of straight travel. To sense their magnetic compass headings, sharks may, but need not swim steadily for a time long enough to have their electroreceptors adapt to the DC component of the electric stimuli they receive by interaction with the earth's magnetic field. It would serve them better to make effective use of the pronounced, highly informative DC component of shorter-duration signals. Their seemingly erratic exploratory accelerations, decelerations, and turning left or right provide much stronger electrical signals about both the ambient electric and the earth's magnetic fields. Moreover, they inform the animal at the same time of prey, obstacles, and boundaries.

Once more, Lohmann & Johnsen try to shove my magnetic orientation experiments under the rug by insinuating that the stingrays might have responded to the transient electric fields induced when the magnetic field was turned on or off. Very strange, because the only biologically valid magnetic orientation experiments in sharks and rays I know of were the ones in which the fields were not "switched" on and off but very slowly "click-free" increased to the desired value and again slowly decreased to ambient again, Kalmijn (1978b, 1981). Even more appalling is Lohmann & Johnsen's reference to the physically ludicrous conclusions of the behavioral experiments with and without magnets attached to stingrays conducted by Walker and his colleagues. These researchers not only take the fabricated attacks of Kirschvink *cum suis* on my magnetic orientation experiments seriously, but also reveal from the physically erroneous interpretation of their own results that they do not understand the application of the motional electric principle as it pertains to their own experiments, thereby denying the animals the use of their own motional electric fields in detecting their magnetic compass headings. Yet, I must with appreciation compliment Johnsen & Lohmann for defining a magnetoreceptor as "a biological structure that can transduce the strength and/or orientation of the local magnetic field to an animal's nervous system." Isn't that exactly what the shark's ampullae of Lorenzini do? Faraday even spoke of "magneto-electric."

Johnsen and Lohmann (2008) furthermore consider the earth's magnetic field weak and require a very sensitive receptor, though weak is a relative statement, as the simplest boy scout compass needle suffices. Moreover, magnetic bacteria are even forced to orient in the earth's magnetic field. The authors also look for microscopic sensors, needles in a hay stack, and complain about the lack of accessory structures such as eardrums and lenses, even though in sharks the whole body is the accessory structure and the sensors are the well-known macroscopic ampullae of Lorenzini (Fig. 4.1).

Johnsen & Lohmann honor me by reproducing an artistic version, on the left, of my original picture of the cranial region of the shark *Scyliorhinus canicula*, on the right, showing the canals and skin pores of the ampullary and lateral-line system. Unfortunately, in Johnsen & Lohmann's picture, the red dots are not just the skin pore of the ampullae of Lorenzini, as their legends say, but include the skin pores of the lateral-line canals as well. In my original, every ampullary pore and every lateral-line canal and pore were copied from the animal I dissected. The small colored dots are the pores of the ampullae, the small open circles are the pores of the lateral-line canals, the solid black lines are the lateral-line canals and the fine colored lines indicate the courses of the ampullary canals. The left image by courtesy of Chris Huh is clearly an unauthorized, mutilated version of my original.

The legends of the cross-section of a shark heading due east, pertaining to Johnsen & Lohmann's version, to the left in Fig. 4.2, show that my original 1981 version, to the right in Fig. 4.2, is not correctly reproduced. Am I to blame for not having been explicit enough? That may well be the case! Below, we will once more

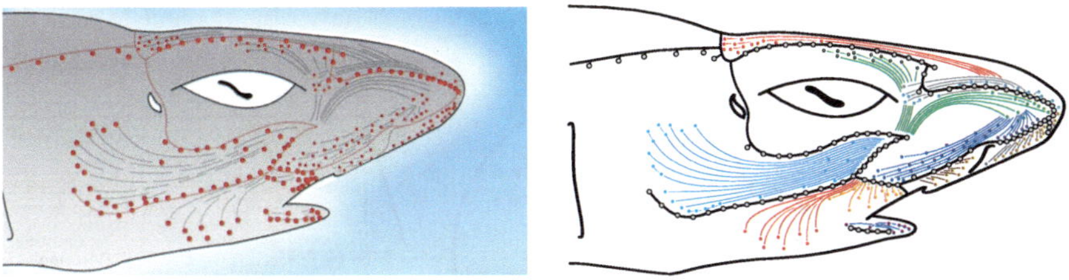

Fig. 4.1 Ampullary and lateral-line sensory systems of *Scyliorhinus canicula*, based on dissection of fresh animal. Left Johnsen and Lohmann (2008), right Kalmijn's original from 1963, 1978. For descriptions, see text

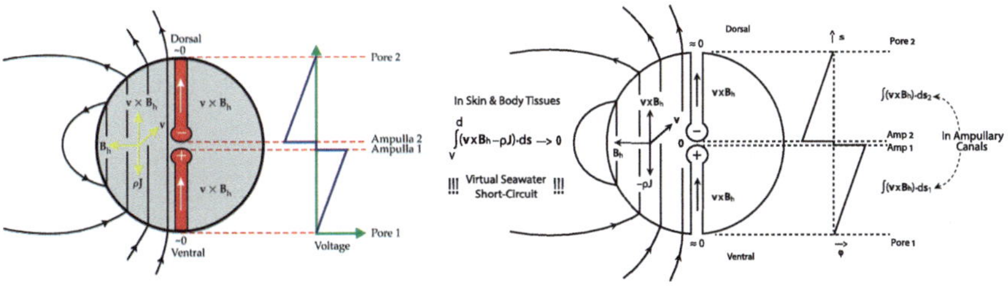

Fig. 4.2 Cross-section of shark head, diagrammatically showing ampullary layout. Left Johnsen and Lohmann (2008), right original by Kalmijn (1981). For descriptions, see text

inquire into the physical intricacies after first identifying the broken links in the chain of Johnsen & Lohmann's reasoning. True, they are not the only confused scientists. I am sorry for them that they ventured to step outside their field of expertise. Why did they not use the old original figures instead of redrawing and corrupting them?

To start, the pores of the ampullary canals are not covered by the skin as they seem to be in their drawing, hopefully by mistake. More serious errors in the text and in Johnsen & Lohmann's figure render their account of my theory incorrect.

As a forewarning of serious trouble, Johnsen and Lohmann (2008) omitted in their figure the minus sign of $-\rho\mathbf{J}$ for whatever reason (Fig. 4.3). Yet, the current density $\mathbf{J}$ and hence $\rho\mathbf{J}$ point in the direction of $\mathbf{v} \times \mathbf{B}$, but $-\rho\mathbf{J}$, the ohmic resistive force, points in the opposite direction, as it should be. Also, the curved lines outside the shark illustrate the current density lines, not the electric field lines. The electric current density lines are continuous across the skin, whereas the electric field lines start and stop at the skin surfaces at different densities, unless the charges on opposite skin surfaces are really nil, which they never are and can-not be, contrary to what Johnsen & Lohmann suppose.

In the legends of the corrupted reproduction of my original drawing, Johnsen and Lohmann state: "*As the shark swims east and into the page with a velocity* $\mathbf{v}$, *its movement across the horizontal component of the Earth's magnetic field* $\mathbf{B}_h$ *causes a vertical electromotive force of magnitude* vB_h. *Because the shark's body and especially its skin are highly resistive, the voltage drop due to the current density* $\rho\mathbf{J}$

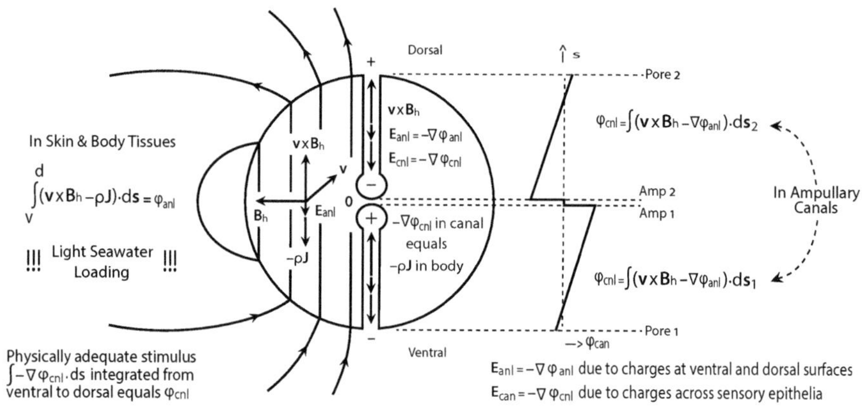

Fig. 4.3 Summary diagram that reduces to the limiting condition of Fig. 4.2 if charge accumulations are considered vanishingly small, though they cannot be zero

results in no potential difference between the dorsal and ventral surfaces of the animal."

In the quoted line, it should read *the ohmic voltage gradient* $-\rho\mathbf{J}$ *due to the induced electric current density* $\mathbf{J}$. You may call $\rho\mathbf{J}$ the voltage drop, but the voltage drop $\rho\mathbf{J}$ is in the direction of the induced current density $\mathbf{J}$, that is, in the direction of $\mathbf{v} \times \mathbf{B}$, not in the opposite direction. In short, Johnsen and Lohmann mutilate my motional electric theory, rendering it fatally flawed, at least their version of it, all because of a missing minus sign, which does seem to be a real mistake, not just a typo. Imagine what happens when, due to a mere change of sign, the direction of gravity reverses.

That $\rho\mathbf{J}$ *causes no potential difference between the dorsal and ventral surfaces* is only true if one hypothetically assumes the potential difference to be zero. But why would it be? Actually, there *must exist a potential difference* due to charges accumulated by the induced electric current on opposite skin surfaces of the animal, even though the charges may be small. The electric field of the charges is what drives the return current through the seawater medium.

But why then is the potential difference across the body *near* zero? Well, because in marine sharks the seawater load carries off most of the charges. When sharks swim into brackish or fresh waters, as some do, their body and skin resistances do not change much, but the increased charge accumulations appearing on opposite sides of the body, needed to drive the return current through the higher resistive aquatic medium, cause the stimulus to be greatly diminished, invalidating the shortcut physical model of *no potential difference* that Johnsen and Lohmann entertain. That is why the original drawing labels the potentials at the pores of an oceanic shark as about zero, ≈ 0. Hence, it is not the resistance of the body plus skin *per se*, but the ratio of the resistance of the skin and body tissues over the resistance of the return path through the aquatic medium, as discussed in my 1981 article they refer to explicitly, which Johnsen and Lohmann apparently did not take notice of.

My original article clearly states that in marine sharks the potential difference *tends to* zero and, in the picture, speaks of a *virtual* seawater short-circuit. However, inside the animal, the always existing electric field **E** of the charge accumulations on opposite skin surfaces, however small they may be, opposes the $\mathbf{v} \times \mathbf{B}$, rendering it less effective. To take this essential issue into account, I have explained that, given $\mathbf{v} \times \mathbf{B} - \rho\mathbf{J} + \mathbf{E} = \mathbf{0}$, it is the line integral of $-\rho\mathbf{J}$, not of $\mathbf{v} \times \mathbf{B}$, that the shark detects. Moreover, the $\mathbf{v} \times \mathbf{B}$ in the canals and the $\mathbf{v} \times \mathbf{B}$ in the skin and body tissues cancel in the measuring circuit of the sensory epithelium, the ampullary canals, the skin, and body tissues, an issue of utmost importance for the comprehensive theory of motional electric orientation. Didn't I point this all out clearly in my very first introduction of the theory in 1974? I am afraid that my pared-down version of the theory for the less physics-trained reader has not been properly understood as the non-realistic limit that the full version tends to if the animals are so heavily loaded by the seawater medium as to be virtually short-circuited, although I have repeatedly warned the reader about the issue.

Paradoxically, as the resistance of the gel-filled ampullary canals increases, the voltage across the electroreceptor membranes also increases, rather than decreases, thus enhancing the stimulus. The reason being that the active negative resistance of the sensory epithelium creates, by positive feedback, a negative canal current enhancing the stimulus, the more so, the higher the canal resistance is, though it must not be too high, Kalmijn (2003) and Bezrukov (2003).

In deference to Johnsen & Lohmann, who tried really hard to find fault with my motional electric research, only revealing their own misapprehensions, let me give a comprehensive clarification of my original 1981 article they consulted. Figure 4.3 depicts the lightly loaded situation of which Fig. 4.2 is the nearly short-circuited version approaching the virtual physical limit, a theoretical limit that can be tended to but cannot be reached. Here, the—perhaps small—but finite ventral and dorsal charge accumulations on opposite skin surfaces are taken into account—as they should—the motional electric $\mathbf{E}_{\mathrm{anl}}$ field of which counters the $\mathbf{v} \times \mathbf{B}$, leaving only the $\mathbf{E}_{\mathrm{cnl}}$ in the ampullary canals due to the stimulus producing charge accumulations on either side of the sensory epithelia to be detected. To repeat, it is the $\mathbf{E}_{\mathrm{anl}}$ that partially offsets the $\mathbf{v} \times \mathbf{B}$, whereas the line integral of $\mathbf{E}_{\mathrm{cnl}}$ is the remaining signal the sensory epithelium detects. It is furthermore clear from the diagram that the pore-to-pore line integral of $\mathbf{E}_{\mathrm{cnl}}$ equals the pore-to-pore line integral of $-\rho\mathbf{J}$ through the skin and body, which the animal detects.

One may look at the situation in two different ways. In the closed measuring circuit formed by the differential pair ampullary sense organs in series and the skin & body tissues, one may argue that in the skin & body tissues the $-\rho\mathbf{J}$ almost nulls the $\mathbf{v} \times \mathbf{B}$, except for the $\mathbf{E}_{\mathrm{anl}}$, so that the ampullae detect the $\mathbf{v} \times \mathbf{B}$, except for the $\mathbf{E}_{\mathrm{anl}}$. That is, the sense epithelia detect the $\mathbf{E}_{\mathrm{cnl}}$ in the ampullary canals, which equals $-\rho\mathbf{J}$ in the skin & body tissues. That is the picture Johnsen & Lohmann apparently had in mind based on my 1981 article. One may also argue, more succinctly, that in the measuring circuit, both the $\mathbf{v} \times \mathbf{B}$ and the $\mathbf{E}_{\mathrm{anl}}$ drop out, leaving only the $-\rho\mathbf{J}$ to detect. The latter version, which I also alluded to as more rigorous in my 1981 article, is preferable as it agrees with the fact that in the animal's frame of reference, the **v** and, hence, the $\mathbf{v} \times \mathbf{B}$ are zero anyhow. The sharks actually will not understand the figures of the arm-chair observer at all, as in the animals' frames of reference there is no $\mathbf{v} \times \mathbf{B}$, neither in the ampullary canals, nor in the skin and body tissues. Instead, there is a $-\mathbf{v} \times \mathbf{B}'$ in the seawater moving with respect to the animal. Although the shark's view in its frame of reference is very different from ours in the earth frame, fortunately for them and for us, it turns out that it is again the $\mathbf{E}'_{cnl}$ which equals the $-\rho\mathbf{J}'$ that the shark in its frame of reference detects.

No wonder that not everybody understands the physical theory of motional electric orientation in sharks and rays. There are many other theories I do not understand either. Maybe, Johnsen and Lohmann should have referred the readers to Kalmijn's (1981) original article they attempt to paraphrase but end up disfiguring. That would have saved me a lot of damage control.

Do not deprive sharks of their electrical sensitivity

Johnsen and Lohmann (2008) further quote for the electrical sensitivity of sharks a scant lower limit of 20 nV/cm, even though I have very clearly proven stingrays to orient in uniform electric fields of 5 nV/cm and inferred small sharks to approach simulated prey from distances at which the field is as weak as 1–2 nV/cm. Other scientists may, in disbelieve, debate those numbers, but the sharks and rays do not, because I learned these numbers from their behavior in the laboratory and in the wild at sea. Furthermore, Johnsen & Lohmann say that "the few experiments that have been done mostly involved training captive animals to respond to the presence of local magnetic field gradients by an electromagnet." True, but why not mention the experiments of Kalmijn (1977, 1978b, 1981), that clearly are not based on "the presence of local magnetic field gradients produced by an electromagnet," but on virtually uniform magnetic fields, simulating those in ocean waters, produced by Helmholtz coils that could be smoothly turned up and down without switching artifacts." In a last attempt to reject the motional electric magnetic sense or sharks and rays, once more, Johnsen & Lohmann quote Walker and his colleagues who concluded that in rays the magnetic orientation was not based on motional induction, proving their total lack of understanding of the physical principles of motional electric orientation pertaining to their alternative experiments.

Lohmann & Johnsen state that "in the several decades since the motional electric hypothesis was first introduced, several findings have emerged that complicate matters." In support of their statement, they repeat Paulin's (1995) mistaken objections, even though these issues were settled decades earlier when Kalmijn presented his motion-electric theory in 1974 when he first introduced the theory. Paulin's and Lohmann & Johnsen's erroneous statement is entirely based on failing thoroughly to study my original papers on the topic. Lohmann & Johnsen choose to rely on Paulin's more recent paper that they apparently did not study or did not understand in detail either, otherwise they would have discovered very soon that it is erroneous. It even advocates the head-swaying mode of swimming as a means of suppressing what they consider *irrelevant* stimuli associated with ocean currents.

Practice what you preach
Lohmann & Johnsen repeatedly boast of how little they know, but that does not mean that little is known. Ironically, they emphasize that "the solution to the long-standing mystery of magneto-reception in animals will almost certainly come from a fascinating interplay of biology and physics." Unfortunately, Lohmann & Johnsen apparently do not constructively take part in that fascinating interplay. To the contrary, till now they proved not to practice what they preached. I very much regret that I felt compelled to articulate those harsh comments on their discussions of the motional electric principle, in particular since Lohmann has done such careful experiments on the orientation of loggerhead turtles.

Lohmann (2010) avoids his earlier stumbling blocks without cleaning up the roadside clutter he and Johnsen left behind. Ever since, they have kept quiet about me, not showing any regrets of their faulty reproduction of my motional electric theory and the unfounded criticism on my experiments. After giving them ample attention in the present article, I hope they will set their record straight.

4.7.2 *Second Troublesome Review Article*

Ignoring their master's voice
Peters et al. (2007) compiled a deceptive prejudiced reinterpretation of the literature on the sensitivity of ampullary electroreceptors, in which I found, to my dismay, several critical issues in need of amendment or forthright repudiation. Since they categorically downgrade the established electrical sensitivity of sharks and rays from 1–2 nV/cm to 20 nV/cm, they throw serious doubt on my original theory of the animals' electric and magnetic orientation. Hence, it is critical for me to scrutinize their scientific credibility and to pass the results on to you, the reader. Anyhow, let us first assess the knowledge of Peters, Eeuwes, and Bretschneider of the kind of electric fields sharks and rays detect, and then discuss their misconceived objections against detection levels lower than 20 nV/cm. It is the very trustworthiness of Peters, Eeuwes, and Bretschneider versus the trustworthiness of the venerable sharks and rays. It would be a failure of the creator, if the electric sense of sharks and rays were electrically not sensitive enough to orient in the ambient electric and earth's magnetic fields in roaming the vast open-ocean waters. Fortunately for the creator, the sharks and rays have proven Peters *cum suis* dead wrong, as I will explain presently.

Physics mastery of Peters, Eeuwes, and Bretschneider
As for the detection of uniform electric fields, Peters *cs* teach us that: "In uniform fields … the gradient can be determined *simply* by measuring the potential at two points, dividing the difference by the distance, and *multiplying* it by the cosine of the angle between the field lines and the line between the measuring points." To understand why this is incorrect, imagine the two points to lie in one and the same isopotential plane which, by definition, the field lines are normal to. Furthermore, lecturing the reader on dipole fields, Peters *cs* give two visually pleasing 3D voltage

plots. However, the plots do not illustrate the physical features of dipole fields that electrically guide sharks and rays to their prey. Instead, they give the voltage distribution with respect to the center of the dipole source, whereas the animals detect the voltage gradients at their position in the dipole field. Moreover, it is not only the strength of the voltage gradient that provides the information, but also, more importantly, its direction. However, their plots do not specifically show either of those. The lines in the plots are the arbitrary grid lines. The field lines and isopotential surfaces, normal to the field lines, would have been more instructive.

Later on, Peters *cs* mention the research of Kajiura and Holland (2002) and make the critical remark that the sharks, in attacking dipole sources at field strengths of 1 nV/cm, may have responded to induction currents caused by motion in the geomagnetic field. It escapes me what they mean by that. But to my surprise, Peters *cs* did not notice the misconception of Kajiura *cs* regarding the configuration of dipole fields and how to use those fields to determine the electrical sensitivity of sharks and rays. To assist the reader in scrutinizing Kajiura articles, let me be more specific. The fundamental cause of deviating from the mathematical physics textbook knowledge is Kajiura's embarrassing error of mistaking the radial directional derivative at a point in the field for the voltage gradient, which it is not in general. This has the consequence that the voltage gradients at points in the normal plane would vanish, instead of being perpendicular to it and being weaker by a factor of two compared to the voltage gradients along the dipole axis, as explicitly stated by Kalmijn (1982), whose formula Kajiura mutilated by multiplying it with a cosine function. This blatant mathematical physics error not only invalidates all inflated sensitivity data reported by Kajiura *cs* based on their misconception of dipole fields, but also makes one seriously doubt about the insights of Peters *cs*, the peer reviewers of their article, and all scientists who have quoted Kajiura concerning the electric fields sharks and rays detect without noticing his fundamental error invalidating all sensitivity data given over the years based on a wrong calibration formula.

Concerning the detection of magnetic fields, Peters, Eeuwes, and Bretschneider state that "the long ampullae of Lorenzini in marine fish can be considered to be part of conducting loops. Turning this loop [in the presence of the earth's magnetic field] will produce induction voltages; even slow rotations can be effective." Such induction voltages could be used directly for navigation." Nonsense! The flaw is that the induced currents upon rotation flow inside the animals in closed loops of which the ampullae of Lorenzini are NOT part as the skin pores open them to the outside (Kalmijn 1984). Hence, by design, rotation *per se* is not providing sharks with useful signals, if they produce detectable signals at all. It is the rectilinear accelerations and decelerations in different compass directions that provide the animals with information of their magnetic headings (Kalmijn 1974).

Electrical sensitivity of sharks and rays not 1–2 nV/cm but 20 nV/cm in prey detection?

In rating the sensitivity of sharks and rays at 20 nV/cm, Peters *cs* refer not only to my 1982 prey detection values of 1–2 nV/cm and Scott Johnson's et al. (1984) close reproduction of them, but also to those of Peters and Everts (1985), who apparently could not ascertain a threshold lower than the 20 nV/cm. Hence, we have to scrutinize their article to find out why they disqualify the tenfold lower results I obtained. Peters and Everts judged the electrical sensitivity based on the responses of the afferent ampullary nerves, as Murray did in 1962, and on my 1962 method of observing the hesitations in the ventilatory rhythm. At the time, Murray arrived at a sensitivity value of 1 μV/cm, while I reported a value of 0.1 μV/cm. Intuitively feeling that I had not reached the ultimate sensitivity values yet, I proceeded by recording the cardiac response to the application of electric fields and found self-evident retardations at values as low as 10 nV/cm. I could have statistically proven the value to be even lower by repeating the test not a few, but many times, but I did not see the merit of doing so at that moment. Those tests were performed in the middle of the night, when the laboratory and road traffic were quiet, and needed rest intervals of an hour between tests. I considered 10 nV/cm already so astoundingly low that it gave me ample room for ascertaining the electric fields in nature that the animals might detect. In short, the best non-behavioral test reported so far provides a conservative sensitivity value of 10 nV/cm, half of Peters' *cs* 20nV/cm, see my 1966 cardiac recordings.

As to Kalmijn 1982 experiments on the shark *Mustelus canis* at sea, the animals performed clear feeding attacks at prey simulating dipole sources from distances where the field strength was 2 nV/cm *or less*. Actually, *or less* was two times less for two of the responses from the largest distance, because the field strengths given in the table were those measured parallel to the dipole axis, while those two responses started normal to the dipole axis, that is, they actually responded at a field strength of 1 nV/cm. My friend Scott Johnson repeated the experiments in a large seawater tank and found a lowest sensitivity value of 5 nV/cm. The remaining difference is no surprise, since my results were from unperturbed, free-ranging sharks foraging at sea under natural conditions, as observed from a rubber raft in the dark of the night, which readily explains the responses to even lower field strengths than those obtained from captive sharks restricted land-based facilities during daytime, whereas sharks prefer to feed at night. However, I would consider an electrical sensitivity of less than 1 nV/cm highly suspicious. At the present state of science, a behavioral threshold lower than 1 nV/cm is beyond our ever-vanishing horizon. Yet, the sharks and rays are the judges, not we.

Concerning the 1–2 nV/cm sensitivity the shark *Mustelus canis* ascertained from the distance of response from a dipole source simulating the bioelectric fields of prey, Peter's *cs* attempts to throw doubt on my gratifying results for reasons that I might not have controlled for obvious alternative explanations, are a cheap and offending remark. Do not forget, I not only discovered the electrically evoked feeding responses but have observed them literally thousands of times in many different shark and ray species over a period of six decades, in the laboratory and at sea. *The detection and approach to a source of electricity is fundamentally different from that in response to an odor field. Thus, in the crucial experiments I conducted in the wild, only those responses were counted where the animals from a measured distance attacked the simulated prey in a single smooth move, without wavering or making corrections when coming nearby. These attacks completely ignored the odor source, placed at some distance from the dipole source, even when during their attacks the sharks had to swim right over it to reach the prey-simulating dipole source.* All experiments were conducted by myself and two more observers, which were instructed to be highly skeptical. Perhaps, as a professional scientist, I was the most passionate trying to conceive alternative explanations to preempt any suspicion of overstatement. In his long list of reliable and highly suspicious threshold values, Peters *cs* make their case against my most reliable threshold data by quoting a value of 19 nV/cm instead of the 1.9 nV/cm in Table 2 of my Science paper Kalmijn (1982).

They should have doubted about themselves, not about the sharks and rays
Peters *cs* say that the extremely low threshold of 1 nV/cm in marine fish has caused wonder and doubt, referring to two articles, Tsong (1994), Petracchi and Cercignani (1998), who, based on single-nerve-fiber recordings, attempted to understand such an astoundingly high electrical sensitivity. As a matter of fact, the authors of the two articles, who I met, did not mean to bring the sharks and rays' performances into dispute but took the challenge to explain their high behaviorally established electrical sensitivity. I have made a great number of nerve recordings myself and agree that they do not fully explain the high sensitivity I established behaviorally, even when the large number of ampullae, nerve fibers, and electroreceptor cells are statistically taken into account. That raises the question: shall the animals be deemed less sensitive to meet our more limited expectations, or shall we have to become more competent to agree with the animals? Peters *cs* even admit having trouble applying and measuring electric fields of such low strengths, which means that they are not qualified to work at these low-electric field levels. They simply will have to accept that sharks and rays surpass them in this respect. Nature is endlessly more refined than our brains can apprehend. Even sharks and rays are not that easy to understand, outperforming us in many respects. Do not doubt about the animals' behavioral performances, just try harder to understand them.

Shot in the foot
Actually, Peters *cs* defeat their own criticism, by saying that turning backwards as the shark passes the dipole source would provide more solid evidence where the dipole source is located via its electric field. That is exactly what I told them beforehand in an invited lecture I gave in their presence in Kalmijn 1995 at the occasion of the 80-year jubilee of our old laboratory in Utrecht, The Netherlands, published in a nice little booklet, where I showed that the turning backward is a fully anticipated behavior based on the guided approach algorithm I have proposed, and showed that the sharks naturally follow the predicted widely curved switchback paths Kalmijn (1995). See also the two photographs of Kalmijn 1978 article, where the small shark swam right over the odor source to attack the dipole source, and where the large shark makes a sharp switchback (Fig. 4.4). Well, the feeding responses are not really part of the present article, but I had to discuss them as they revealed the highest electrical sensitivity of the animals, unaware of being objects of investigation and under the most natural conditions. Surprisingly, in the abstract Peters *cs* say "symmetrical stimulation of the sensory raster would provide an easy

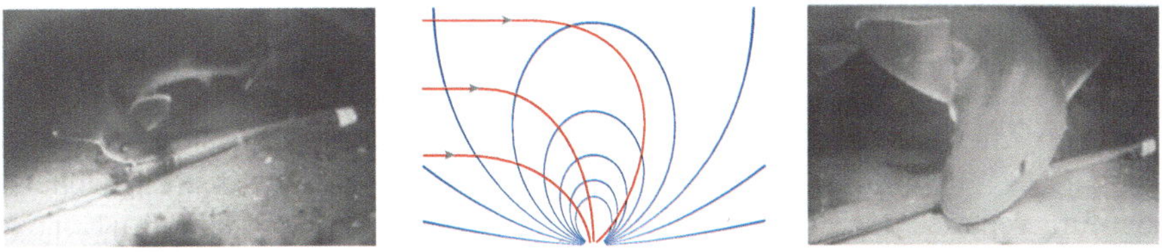

Fig. 4.4 Left: Juvenile *Mustelus canis* swims over odor source (at white marker) to reach and bite dipole source simulating prey. Right: Adult *Mustelus canis* makes sharp switchback to attack dipole source. Middle: Approach paths (in red) traced by Kalmijn's DC bioelectric-field algorithm, including switchback

way to keep a straight course with respect to a far-field stimulus. The same neural mechanism would be effective in the location of a bioelectric prey generating a near-field stimulus." These are just as hollow statements as if somebody said, "the animals just go where the source of the prey's electric field is." If it is that simple, then my orientation and prey-detecting researches will have been a waste of time.

Can sharks and rays orient in uniform fields of 5 nV/cm, or in fields of 20 nV/cm at best?

Peters *cs* furthermore try to question my behavioral experiments that proved elasmobranch fishes, exemplified by the stingray *Urolophus halleri,* can orient in uniform electric fields of 5 nV/cm and in a horizontal magnetic field of 26 μT. In 1978, I published an initial report on the stringent protocol and the first results of the *magnetic-field* orientation experiments. In 1981, I elaborated on the use of the appropriate method of sequential analysis applied and published the complete *magnetic* data set. In Kalmijn 1982, I also reported on my new *electric-field* orientation experiments and gave the complete *electric* data set. In Kalmijn 1984, I reviewed all the data once more. Why Peters *cs* say that the experiments lacked a proper methodological description, escapes me altogether. What more could I have told the reader? These were the most demanding and most thorough long-term conditioning experiments I have ever conducted. The electric and magnetic experimental setup was specifically built for these experiments in a remote, electromagnetically quiet region of the Quissett Campus of the Woods Hole Oceanographic Institution. The most convincing data were obtained by two summer students who honestly were loath to believe in my results. Yet, they conducted one full series of experiments to see for themselves, which turned out the most significant one of all.

Let me counter a few more mistaken remarks of Peters *cs*. They criticize me for concurrently using three stingrays, simply to speed up the experiments, without telling how this influenced the response criterion. Well, as a biologist, I knew by experience that single rays in isolation do not perform as well as in groups of three. As for the response criteria, per trial, I only allowed one of the three rays to enter the enclosures, whether the correct one or the incorrect one. After that the gates of both enclosures were closed. As to how I rewarded the ray that entered the correct enclosure, well I gave it a small piece of herring as soon as it had entered, and the gate was closed. If the ray entered the wrong enclosure, it was chased out gently by prodding it lightly with a smooth plastic rod. How long did the experiments take? That depended on the motivation of the animals. After 10–20 runs, the animals tended to slow down and the session was terminated, which usually took about half an hour. After each trial the stingray captured in the enclosure was set free after moving the enclosure randomly to either the north or the south of their habitat. I dutifully reported this all in my articles. See, for instance Kalmijn (1978b). None of this was of importance for the statistical method I had chosen. It all is simply a matter of clean experimentation and faithful reporting. Moreover, the results were so robust that the most critical observers, performers as well as visitors, had not the slightest doubt that indeed the animals were orienting in the electric and magnetic fields respectively. Just because the experiments went so well, I and my dear coworker, my wife Vera, were very concerned not to be deceived. Neither we, nor any skeptical people we invited, could think of explanations other than that the animals acted in response to the electric and magnetic fields applied. Hence, I strongly believe in what my favorite animals revealed to me.

Need and choice of statistics

Peters *cs* consider the 66% of correct choices small compared to those of the psychophysical studies that usually ask for 70 to 75%. Yet, mine were not psychophysical studies to determine whether the animals could associate ambient electric and magnetic fields with food. They were designed to prove the animals' ability to detect the ambient electric and magnetic fields and make biologically significant use of them. They were not meant to determine thresholds, but to establish sensory abilities at stimulus strengths encountered in ocean waters. Of course, just the percentage of correct responses is not a criterion of significance. It is the small likelihood of getting those percentages based on random choices calculated by rigorous statistical mathematical procedures, for which I applied the powerful method of sequential analysis. My experiments prove that the results are more likely, at α and β errors of 0.001 or less, to be due to a 70–75% correct ratio than to a 50% correct ratio. Dear reader, mind your statistics, read my original articles and the textbook by Dixon and Massey (1969) and the monograph by Wald (1947) on which my statistical evaluation was based and judge for yourself. Mine were not 2AFC tests as Peters et al. (2007) call them. The animals were not forced to make choices but were free to enter either one or neither one of the two enclosures. Why trying to talk away the obvious by unjustly criticizing my statistics?

That in 1966 I did not apply any statistics to the results is simply because the results were self-evident and reproducible on command, any time. Do you need to be hit on the nose ten times before you know statistically for sure that you were hit on the nose? I apply statistics that I believe after a thorough study to be the most suitable for my experiments. See Kalmijn (1981). That is the responsibility of a scientist. Certainly, for Dijkgraaf our teacher, self-evident was more convincing than statistically proven. As for the stingray experiments, any experienced animal naturalist would, by witnessing my experiments, have read from the animals' deliberate behavior that they "know" what they are doing.

Having said this all, one may wonder why Peters *cs* degraded my values of 1–2 nV/cm in prey detection, of 5 nV/cm in ambient-electric field detection, and of 10 nV/cm by the 1966 heartbeat decelerations recordings, to their 20 nV/cm. The only answer I can think of is: just because they trusted their recordings from the nerves and their observations of the respiratory reflex more than the behavioral orientation of stingrays in electric fields of natural configuration and strength, and from the behavioral predatory responses of free-moving sharks in the wild. If they would have qualms with the electrical prey detection experiments, they should at least have reported the very sure level of 5 nV/cm that I learned in the uniform orientation experiments. Do not forget though, the 1–2 nV/cm prey detection experiments were carried out in the wild in the shallow open waters of Martha's Vineyard off Cape Cod on foraging sharks in the dark of the night, in the presence of the undisturbed ambient electric and magnetic fields, as natural as it can be.

In the same paper, Peters *cs* were even speaking of the animals' frequency response determined by applying single sine waves in the water, not realizing that single sine waves do not consist of single frequencies. As for the peripheral detection threshold, I must remark that in terms of detectability, that is, in terms of signal-to-noise, the frequency response of the ampullae of Lorenzini is flat from about 4 Hz down to 1/8 Hz, the lowest frequency tested, or less. I mean, at 1/8 Hz there is still no significant signal-to-noise decline in the response. Admittedly, this depends on how the action-potential trains are statistically processed, but I may assume that I am not any better at it than the sharks and rays are. My results are in terms of frequencies well in line with the behavioral response. Between the ampullary sense organs and the central nervous system, differentiating and integrating steps are likely to occur, affecting the frequency response at the various stages in the detection process. Moreover, what really counts is the frequency response of the *sensory system*, including the central processing and the exploratory behavioral movements.

Do not tell sharks how motional-electrically to detect the earth's magnetic field, they know better

Upfront in the abstract, Peters *cs* state: "Angular movements are essential for stimulation of the ampullary system in uniform d.c. fields." I agree. Furthermore, I quote: "Angular movements in the geomagnetic field also generate induction voltages [gradients they mean], which exceed the 20 nV/cm limit in marine fish." The latter statement is of dubious import, as it is based on a misunderstanding of the physics involved. In the first place, the ampullae of Lorenzini are not part of conducting loops as Peters *cs* say, probably misled by Paulin (1995) and perpetuated by his followers. The induced current actually flows across the body bypassing the ampullary sense organs. The sense organs, acting as voltmeters, measure the voltage drop caused by the motionally induced currents flowing across the body (Kalmijn 1974). In the second place, angular movements in uniform magnetic fields are as such ineffective in stimulating the ampullae of Lorenzini of sharks and rays, as the induced currents circulate in and around the body rather than traversing it as is necessary for the detection of the magnetic field as explained in Kalmijn (1984). The adequate stimuli are not the turning signals *per se*, but the forward motions in different directions. Those indeed exceed 20 nV/cm at linear forward speeds of 10 cm/s in magnetic fields of 20 μT, as I have made clear from the beginning, but they exceed 2 nV/cm even at speeds as low as 1 cm/s.

The elasmobranch ampullary sense organs acting like a stochastic resonance system, as Peters *cs* infer, will not resolve their fundamental misunderstandings. The reader deserves more clarity and evidence before accepting stochastic resonance as a magic cure to raise the electrical sensitivity 20-fold, even if true. As of yet, the statement is utterly unfounded. There are several other statements that I do not comment on, bizarre concepts of gain curves of Bode plots, near field vs far field, etc., but that does not mean that I agree with them. But enough is enough.

4.8 Per Se, Direct, and Motional Electric Magnetic Detection

After reading Kalmijn (1982), Meyer et al. (2005) brazenly state that "to date, there is no empirical evidence that sharks can detect magnetic fields *per se*," whatever they mean by *per se*: not just the nonuniformities in the magnetic field but the magnetic field as such? In 1978, I had already reported the leopard shark, *Triakis semifasciata*, to avoid a magnetic field produced by an 8″ in diameter coil held outside the wall of the shallow fiberglass observation tank, producing a local magnetic field measuring merely 25% of earth strength, as well as to execute apparent feeding responses directed at a small bar magnet held under the bottom of the tank, even though the magnetic fields remained constant during the observations. Furthermore, I noted that the animals changed their habitual northern early-morning rest locations in their 20 ft in diameter outdoor habitat, until I nulled the vertical component of the magnetic field with a large coil wound around the observation tank, which caused the animals to distribute in a random fashion. All showing evidence that sharks detect magnetic fields, though not yet proving that they have a magnetic sense. After describing these circumstantial evidence providing observations, I published the first definite data, obtained from my overly scrupulous conditioning experiments, proving the stingray *Urolophus halleri* not only able to orient biologically meaningfully in ambient electric fields of 5 nV/cm, Kalmijn (1982), but also in ambient magnetic fields of earth strength, Kalmijn (1978b, 1981). *These latter magnetic experiments are the only behavioral experiments I am aware of that prove the animals not merely to detect magnetic fields but to have a genuine magnetic sense, as they performed a perfectly natural behavior guided by a perfectly normal earth's magnetic field.*

That Meyer, Holland, and Papastamatiou (2005) were not aware of, at least did not refer to, my earlier magnetic orientation experiments baffles me, especially since Kalmijn (1978b, 1981) were listed in my 1982 article, which they apparently had consulted. Or did Meyer, Holland, and Papastamatiou (2005) discard my experiments, because they just were not responses to magnetic fields *per se*? However, compared to my original, biologically meaningful experiments, Meyer,

Holland, and Papastamatiou merely observed responses to violent magnetic seaquakes. Nevertheless, they consider their method "a robust behavioural assay that can now be used to precisely determine how sharks detect magnetic fields and to measure detection thresholds." Unfortunately though, their hope and expectation did not work out that way in subsequent experiments, as would become apparent from Anderson et al. (2017).

Another restrictive term often used is "*direct*" versus "*indirect*" magnetic detection, by which certain particular scientists seem to mean "any magnetic detection mechanism other than by motional electric induction," an often tacitly implied artifice to interpret the literature selectively, in order for the "*direct magnetic*" scientists to deal with their own limited success in contrast with the successful "*indirect magnetic*" scientist. They may just as well say "by any other means than Kalmijn's motional electric detection of the earth's magnetic field in sharks and rays." Let it not be the ampullae of Lorenzini. That is too self-evident. Let it be magnetite after all, like the bacterial magnetite Blakemore (1975) discovered and of which Kalmijn and Blakemore (1978) determined the magnetic moments orienting the organisms in a biologically meaningful fashion. True, the two are not mutually exclusive. Nature is redundant, taking advantage of all possibilities.

As a matter of fact, physically electric fields do not act directly on electric fields nor do magnetic fields directly act on magnetic fields, according to the principle of superposition. Magnetic fields may act on the sources of magnetic fields, but on magnetic fields *per se*. Therefore, the measurement of electric and magnetic fields is always *indirect*. Well, let us first agree, if an animal orients in the earth's magnetic field, it has a magnetic sense, whatever its detection mechanism may be. So, what is the cardinal difference between direct and indirect? Good question. Well, how is the earth's magnetic field generated? Textbooks tell us: by motionally induced currents in the earth's core. Even the magnetic field of magnetite results from circular electric currents, if we believe Ampère. What then is the most direct indirect detection of magnetic fields? Just guess: by measuring the motionally induced electric currents magnetic fields produce. Einstein could explain that without using the term magnetic. The terms are merely a matter of a human convention. Electric and magnetic are just two manifestations of one and the same electromagnetism. Animals leisurely practice what many scientists have difficulty understanding.

Walker et al. (2003) go even so far as to redefine a magnetic sense by limiting it to responding only directly to magnetic fields stimuli, not by extracting magnetic field information from interactions of the magnetic field with other specialized sensory systems. To corroborate their restrictive exclusion principle, they falsely claim that their impairment studies strongly imply that magnetic field detection in elasmobranchs is also based on single-domain particles of magnetite. This prejudiced statement would lead one to believe that sharks and rays, who have behaviorally been proven to orient to the earth's magnetic field, would not have a magnetic sense if they detect the earth's magnetic field motion-electrically with the ampullae of Lorenzini, but would have a magnetic sense if they detect the earth's magnetic field based upon an unproven single-domain magnetite basis. Too bad, the conclusions drawn from Walker, Diebel, and Kirschvink's behavioral impairment experiments that their animals did not orient motional electrically were not based on proper physical reasoning, as I have explained above. They apparently want to get sharks and rays also in their own lofty magnetite castle, not in the well-established court of my trusted teachers Faraday and Einstein. I prefer to leave it to the sharks and rays to decide.

4.8.1 Legacy of Paulin

To analyze Paulin's left-over crumbs

As a sequel to Paulin's (1995) article, Molteno and Kennedy (2009) present a mathematical analysis of their university colleague's induction-based magnetoreception model in elasmobranch fishes, relying on Paulin's left-over crumbs, an analysis specifically recommended by neurophysiologist Prof. D. Walter Stühmer at the Max Planck Institute, Germany. As to the functioning of the ampullae, Molteno & Kennedy implicitly abide by, rather than question, Paulin assuming the ampullae to detect the motionally induced currents in the ampullae of Lorenzini, whereas neurophysiologists, including myself, have proven the induced electric currents, after a very brief transient, to bypass the very high resistance ampullary sense organs altogether, flowing instead across the animal through the skin and body tissues. The mistakenly inferred motionally induced ampullary currents invalidate Molteno & Kennedy's mathematical introduction, as they base their argument on the Lorentz force, ignoring the ohmic resistive force acting on the ionic charge carriers moving in the ampullary canals, at least according to Paulin who Molteno & Kennedy devotedly believe in. To their defense, Molteno & Kennedy are not the first scientists to "explain" the motional electric magnetic sense, saying "oh, that is simple, it is just the Lorentz force." It is not at all that simple, though.

Gross misinformation

This is not the only instance of Molteno & Kennedy's serious lack of fact-based knowledge. In reviewing the literature concerning the detection of the geomagnetic field, they refer to behavioral experiments by Murray (1962) and Andrianov et al. (1974), even though neither of them report on any behavioral experiments they conducted; to Meyer et al. (2005), who used a highly offending magnetic sledgehammer technique, even though not Meyer et al., but Kalmijn in 1978 proved the animals indeed capable of orienting in a steady uniform magnetic field of earth strength after, in 1974, offering the first physically correct induction-based magnetic orientation theory. Even worse, Molteno and Kennedy state that I merely suggested the compass sense, but Paulin (1995) refined it, rather than realizing that Paulin made a futile attempt to destroy my original theory and came up with a fatally flawed plagiarized version of it.

The Lorentz force misses the crucial point

To learn from the often-made mistake of calling upon the Lorentz force, see Lorentz (1892), let us consider the ruling physics more closely. Molteno & Kennedy start out stating that, in the ampullary system, the **E** and $\mathbf{v} \times \mathbf{B}$ of the Lorentz force

$\mathbf{F} = \mathbf{E} + \mathbf{v} \times \mathbf{B}$ are in equilibrium, hence $\mathbf{E} = -\mathbf{v} \times \mathbf{B}$. The Lorentz force is the *electrical* force acting on charges, but in fluid and solid media the particles carrying the charges are subject to *non-electric* forces as well, to wit, in conductors to the ohmic drag forces $-\rho\mathbf{J}$ and in non-conductors to the polarization forces $-\mathbf{P}/\varepsilon\chi$. Neither of these two forces are included in the Lorentz force, not by neglect but by design. Hence, since in Paulin's model, which Molteno & Kennedy adhere to, the charges do move, they should have included resistive force $-\rho\mathbf{J}$ in the equilibrium equation for the ampullary canal: $\mathbf{F} = \mathbf{E} + \mathbf{v} \times \mathbf{B} - \rho\mathbf{J}$. To their defense, Molteno & Kennedy may have assumed that the sensory epithelium does not allow ions to pass, but then they should have explicitly stated so, because Paulin emphatically assumes the excitation current to be funneled through the ampullary canals, not through the skin and body tissues. Ironically, yes, there is a current flowing in the ampullary canal, but not Paulin's induced current, but a current in the opposite direction actively produced by the receptor epithelium behaving as a negative resistance. So there is a $-\rho\mathbf{J}$ anyhow, rendering the Lorenz force incomplete for Molteno & Kennedy to base the paper on.

The animals told me otherwise

Next, Molteno & Kennedy (2009) compare the $\mathbf{v} \times \mathbf{B}$ compensating electrostatic field $\mathbf{E}$ in the ampullary gel with the threshold strength of the electric fields that the animals respond to behaviorally. They therefore review the electrical sensitivity of sharks, misquoting the sensitivity in the smooth dogfish, *Mustelus canis*, which I reported as 1–2 nV/cm, not 1–2 μV/m or 10–20 nV/cm, hence off by a factor of ten. They, furthermore, quote the electrical sensitivity of the hammerhead shark, *Sphyrna lewini*, and the sandbar shark, *Carcharhinus plumbeus*, determined by Kajiura and Holland (2002) to be 0.05 μV/m or 0.5 nV/cm apparently not aware that the calibration formula Kajiura & Holland concocted, by modifying the correct one I had given in 1982, was *DEAD WRONG*. As a result of the latter, they fell into the trap of not only overestimating the animals' electrical sensitivity, but also having the wrong configuration of the dipolar test fields in mind. See Scholium. I believe that the animals know better, at least, they behave as though they do. Referring to Tricas and New (1998), Molteno & Kennedy quote for the frequency response of the afferent nerve fibers a range from 0.1 to 10 Hz, even though the original authors give actual −3 dB cutoff values of 0.5–4 Hz and the animals' *behavioral* responses show much greater sensitivity at the lower frequencies than those numbers suggest. It is not just the peripheral receptors but the entire sensory system, including the exploratory movements that determine the behavioral frequency response of the electric sense.

The numbers prove their mathematical analysis inaccurate

With this straightened out, let us continue. Molteno & Kennedy (2009) develop the swerving stimulus in terms of its frequency components, following Paulin's approach, but instead of talking about the distortions in the waveform received as compared to a pure sinusoid or to the signals received from the semicircular canals, Molteno & Kennedy express the distortions in terms of harmonics. Let us look at the results. At a swimming speed v of 1 m/s, a horizontal magnetic field of 25 μT, and an angular modulation α of 0.5 rad, they give 23.4 $\sin\theta$ μV/m for the DC, 12.1

cosθ μV/m for the first harmonic, and 1.53 sinθ μV/m for the second harmonic. Hence, the harmonics range from zero to the full amplitudes given. Molteno & Kennedy conclude these amplitude values to be well above a threshold of 0.1 μV/m or 1 nV/cm, of course, except when either sinθ or cosθ are close to zero. The ratio between the first and second harmonic would give the shark its compass heading, and by keeping this ratio constant the animal would be able to maintain a constant heading. Simple as pi.

However, if we apply Molteno & Kennedy's reasoning to sharks traveling at a more natural speed and swerving angle, that is, at a v of 0.5 m/s, rather than a rather large 1.0 m/s and an angular modulation α of 0.1 rad, rather than the wildly exaggerated 0.5, I calculate at a horizontal magnetic field of 25 μT for the first harmonic 1.25 cosθ μV/m (from 12.5 to 0 nV/cm) and for the second harmonic 0.03 sinθ μV/m (from 0 to 0.3 nV/cm). Hence, to keep a constant ratio between the first and second harmonic is out of question, unless the animal swerves like mad for a while. Moreover, the second harmonics that Paulin built his new theory on are largely determined by the choice of simplifying assumptions made to render the ampullary model tractable, hence artifacts. And why would a shark try to read his magnetic compass heading based on the ratio of harmonics, if a couple of individual accelerating and turning movements are not only sufficient but much more powerful to determine both the direction and strength of the ambient electric as well the earth's magnetic field, making use of the very DC component that Paulin, Molteno, and Kennedy reject, rather than have the animals wait until the DC at the receptor level has died out and several cycles have passed to do the frequency analysis? The analysis of Molteno & Kennedy is interesting but far out.

Time domain versus frequency domain

Does the shark's task of electric and magnetic orientation really become more convenient or more precise by analyzing the stimulus in the frequency domain, even if rectilinear accelerations and individual turns suffice? Meaningful frequency-domain data become available only after the initial transient has vanished, after adaptation of the sense organs and several cycles have passed. To the contrary, time-domain, instead of frequency-domain methods, use the data right away as they come in, preferably even before significant adaptation of the sense organs takes place. Time domain results can still be improved by averaging independent, repeated determinations. Certainly, in prey detection the system had better operate in the time domain, or else the animals would starve to death. The guided approach to prey is possible only if the shark can take the momentary ambient electric and magnetic fields into account, which from their behavior it is evident that they do.

Physics blunder perpetuated

Molteno and Kennedy (2009) next scrutinize the condemning statement made by Kirschvink et al. (2001) that a strong permanent magnet inserted in one of a stingray's nares would not affect an induction-based magnetic orientation mechanism, because the animal's electroreceptors would not move relative to the magnet. Molteno & Kennedy object, however, that the ray's body is not perfectly rigid and, hence, the electroreceptors may move slightly with respect to the magnet anyhow, though these motions will be very small in the animal's rather rigid head. In saying this, they apparently overlook the obvious cause of a highly offending motionally induced electric currents of overriding importance: the current induced by the seawater moving relative to the permanent magnet. I repeat: Imagine the magnet to be oriented along the shark's left-right, transverse body axis. The magnetic field directly above and below the animal will then be directed horizontally. By interaction with the water through which the shark and the attached magnet move, a vertical electric current will be generated, stimulating the animal along its ventro-dorsal axis, the axis along which it is supposed to detect the induced electric field resulting from its forward motion in the presence of the horizontal component of the earth's magnetic field. In short, they made again the old mistake of Lissmann and Machin (1958) perpetuated by Kirschvink et al. (2001), to consider the local $\mathbf{v} \times \mathbf{B}$ at the position of the sense organs to be the physically adequate stimulus, rather than the local $-\rho\mathbf{J}$ resulting from the $\mathbf{v} \times \mathbf{B}$ and $\mathbf{E}$ throughout the whole seawater volume moving with respect to the animal.

Repeat performance

More recently, Kirschvink et al. (2010) repeated their physically fallacious claim in even stronger terms: *"Recent work has even shown that marine animals such as sharks and rays—which do have a highly developed electroreceptor system with macroscopic structures called the ampullae of Lorenzini—do not appear to use it to detect the geomagnetic fields as long has been suggested."* This surprising discovery came from the erroneously interpreted experiment of fixing a permanent magnet to the snout of stingrays. See for their physically nonsensical reasoning the foregoing paragraphs pertaining to the authors' 2001 and 2003 articles, and my comments on Molteno and Kennedy's (2009) article. Kirschvink, Winklhofer, and Walker (2010) admit that *it can be difficult to understand and apply the theory and approaches of multiple disciplines effectively and to analyze and interpret well the results of experiments that cross disciplinary boundaries.*

Sharks responding to magnetic seaquakes, continued

The influence of the scientific ignorami is frightening. Anderson et al. (2017) put my theory of motion-electric magnetic detection by sharks in doubt and have Paulin rescue it, not realizing their doubts are based on not understanding my theory and Paulins' rescue is based on a distortion of it. Well, after reading their present article, it becomes evident how little the authors understand not only what I and Paulin but also of what they themselves are doing.

Describing their reproduction and extension of Meyer, Holland, and Papastamatiou's original 2005 experiments, Anderson et al. (2017) state that "Activating the artificial [ambient magnetic] field produced an *immediate* response in the conditioned sharks. They changed from swimming steadily around the perimeter of the tank to swimming faster, turning rapidly and converging on the target [area in the center of the tank] in anticipation of a food reward. This change in behavior resulted in significantly more passes over the target with the artificial field on than off [at strengths from 0.03 to 2.89 μT]." The authors show this behavior convincingly, as did Meyer et al. (2005) at field strengths ranging from 25 to 100 μT. Anderson et al. (2017) meant to determine what it was that the sharks did and did not respond to, a laudable effort. Furthermore, they "attempted to induce magnetosensory impairment in sharks by reversibly attaching [neodymium] magnets to the dorsal surface of the head. The placement of magnets on the head was designed to impair magnetic stimulus perception." "The attachment of the magnets was judged not to impair the conditioned feeding responses." Unfortunately, in my opinion, their experiments created more problems than they solved. Yet, I tried very hard to make sense of their well-meant effort. Please, correct me when and where I am mistaken in the following commentary.

Anderson et al. (2017) missed one great opportunity, as did Meyer et al. (2005). Why did they not switch the magnetic field on and off using a "click-free switch" by first slowly and smoothly turning the magnetic field up and later slowly and smoothly down again, as Kalmijn did in his 1978 experiments? This would have prevented the electromagnetic switching artifact and made much of the already quite dubious paper superfluous. At least, they try to quantify the electrical switching transient but, unfortunately, fail to do so correctly. They start off with Faraday's Law, equation (1) $\varepsilon_{ind} = \int \mathbf{E}_{ind} . d\mathbf{s} = d/dt(\Phi_B)$, where they call ε_{ind} the induced electrical field in "Volts m" [*sic*] instead of the line integral of the induced electric field, the emf, in "volts." Moreover, the emf is not produced "at points" but "between points." In equation (3), they consider the magnetic field to be uniform within the coil, as if the windings formed a long cylindrical coil. Even more surprisingly, they apparently were not able properly to measure the pertinent electric fields in the experimental tank, belying the title of the article: "Insight into shark magnetic field perception." If the electrical background fields were that bad in their large observation tank as they measured them, however they did so, well, then their tank was not suitable for electric and magnetic experiments at the level at which sharks operate their electric sense.

How to interpret the results of Anderson, Clegg, Veras, and Holland? Admittedly, I had trouble following their logic and may have understood it wrongly at some places. I apologize for that. As for their physics proficiency, they repeatedly use erroneous units: "*induced electrical field (Volts m)*" and "*milliVolts per meter (mV/m*$^{-1}$*)*" and "*nanoVolts (nV) cm/s*$^{-1}$, and "*1.5 v applied charge*", and "a *voltage gradient of 74.35 nV cm/s*$^{-1}$", and "a voltage of 35 nV cm^{-1}, unless they all are misprints. However, a background electrical flux of "*1 mV/cm/s*$^{-1}$" is certainly not a misprint, but just nonsense. Electrical flux is through an area and is measured in volt·m. Dimensions and scientific terms wrong ==> all wrong. Yet, much worse than those petty details, they refer to Paulin (1995) as the authority on motional electric orientation in the earth's magnetic field, even though he neither understood the physics nor the biology of motional electric orientation, as I have explained above. They also quote an electric sensitivity of 0.5 nV/cm given by Kajiura and Holland (2002), not aware that the sensitivity calculations by them were dead wrong, as they were based on a physically nonsensical calibration formula, a corrupted version of my 1982 one. Furthermore, Anderson et al. apparently do not value Kalmijn's 5 nV/cm undeniable ambient-electric field sensitivity data and his 1–2 nV/cm bioelectric prey field sensitivity data, for their own sake preferring the lesser sensitivity of, on the average, 35 nV/cm.

The most puzzling of Anderson, Clegg, Veras, and Holland's results is the fact that (1) the responses were equally robust whether the stimulus strength was on the order of either 3 μT or 0.03 μT, that is about ten to a thousand times smaller than the local strength of the earth's magnetic field, yet still with their brief calculated electrical transients well above the sensitivity level of the sharks, to wit, 73.3–7300 nV/cm, as given in their Fig. 2.4, and (2) that the strong magnets on the head of the animal affected, but did not eliminate the trained performance. I can think of many control experiments that would and should have made the results more credible. It is clear to me that more factors played a role than the authors were aware of. The paper was simply published prematurely.

True, the electrical transients, though strong, were very short, beyond the low-frequency range of the shark's electric sense. Too bad that they were not prevented to begin with by use of a "click-free switch." Well, how then, without the transient, would the magnetic field step effect the shark's electric sense? Let a shark swim at a velocity of 0.5 m/s in an experimental tank without ambient electric field in the presence of a magnetic field of which a vertical component is 20 μT. The transverse electric field it then receives is 100 nV/cm. The ampullae of Lorenzini adapt to the signal in a matter of a few seconds. Anyhow, when suddenly the magnetic field changes by, say 10%, then the animal receives a change in the transverse field of 10 nV/cm, still within its sensitivity range. However, when the field changes by merely 0.1%, the animal receives a change in the transverse field of 0.1 nV/cm, well below its sensitivity range. Hence, at the low magnetic stimuli applied, the electric sense of the shark does not explain the detection of the magnetic field change.

Even though when a shark senses a transverse electric field in response to a change in the vertical component of the ambient magnetic field, it would not tell the animal anything about magnetic field detection, because such a sudden change would be the same if suddenly a horizontal electric field would appear in the experimental tank. There is no reason for the shark to interpret this as due to a change in magnetic field. It would more likely appear to the animal as due to the presence of prey. Whether in general sharks interpret the electrical signals they detect as what we call either electric or magnetic depends on what for the animals the signals based on innate or learned experience represent, taking its exploratory movements into account. The animals are oblivious to how we make the stimulus.

What this all indicates to me is that neither the electrical switching artifact nor the motionally detected change in the vertical component of the magnetic field plausibly explain the animals' responses in Anderson, Clegg, Veras, and Holland's training experiments, as the responses remain robust down to levels at which sharks certainly do not operate the electric sense anymore. Moreover, the magnets attached to the head would not only interfere with a magnetite sensor, but just as well with a motional-electric sensor, as I have explained above. What else it might have been, I would not be able to tell or suggest, unless I would be personally invited to witness and scrutinize the experiments. I am most surprised that the responses of unimpaired animals are essentially the same from 100 μT in 2005 to 0.03 μT in 2017, as if the strength of the magnetic stimulus is irrelevant, which gives me an eerie feeling, unless it is something else the animals were trained to respond to. That the reactions of the animals were less vigorous after attaching the magnets only after lowering the magnetic stimulus is an interesting observation.

In short, I failed to arrive at a coherent picture from what the authors tell us. Certainly, their electric and magnetic background and their knowledge of electrical equipment is deficient. Yet, my suspicion is that there were several uncontrolled factors interfering with the experiments. Hence, I can only guess which of their results are facts and which are artifacts. How sensitive the animals can be to all kinds of spurious stimuli became clear to me when executing the electric and magnetic experiments in the stingray ray *Urolophus halleri*. The small magnetic test facility was located in the quiet woods of the Woods Hole Quissett campus. When in the morning we quietly walked down the sloping leafy trail to the facility to conduct our daily trials, the stingrays were commonly found slowly to swim around with sand on their backs, clearly from just emerging from hiding in the sand. Obviously, our footsteps during walking down the path had alerted the animals of the upcoming breakfast. In my present Electromagnetic Research Facility, the observations tank is founded on the bedrock, isolated from the platform on which we are moving around the tank.

Biological artifacts

In their experiments on the magnetic sense of sharks, Meyer et al. (2005) resorted to an electrical sledgehammer method by switching on a vertically directed magnetic field of 100 μT, which produced a strong electrical transient in the shark tank, upon which the animals responded immediately. No wonder, merely read the article of Anderson et al. (2017) for more detailed information before judging: the latter speak of a transient of 73.3 nV/cm where the animals swam at the periphery of the tank at a magnetic field stimulus of 0.029 μT, that means, at the whopping 100 μT applied by Meyer, Holland, and Papastamatiou as much as 253,000 nV/cm, whereas the sensitivity threshold must have been at least as low as 5 nV/cm. Besides the well-documented induction-based detection of magnetic fields, the authors mention magnetite-based receptors and magnetically sensitive chemical reactions, even though there is no credible evidence for those to be used by sharks and rays. But, you never know, certainly not if, concerning the motional electric principle practiced by sharks and rays, you are not aware of the sharp contrast between the physics and biology of the present author and that of Paulin (1995).

More biological artifacts, dodging the issues

Newton and Kajiura's (2017) article on magnetic field discrimination, learning, and memory in the yellow stingray elicits some of my early-day reminiscences. Lissmann 1958 wrote: "*Gymnarchus* can be pulled along, as it were, by a magnet which is slowly displaced along the slate wall of its aquarium. The fish also has been trained to avoid its feeding place and food when a stationary magnet was placed near it (not visible), and to accept food when the magnet was removed, and vice versa." Lissmann, though, was smart enough not to consider his observations as evidence for a magnetic sense. Inspired by Lissmann, I also tested the responses of sharks and rays to bar magnets hidden in the bottom sand of the observation tank in the early days and interpreted the results as an indication of an electrical sensitivity. Even rusty nails and air stones with inside a higher or lower salt concentration than the surrounding seawater were attacked. Yet, I did not publish these trivial, biologically meaningless observations, but proceeded in earnest with testing the animals' behavioral responses to the bioelectric fields of prey, to uniform electric field simulating the common fields of ocean streams, and to uniform magnetic fields faithfully mimicking the earth's magnetic field. Their inference that my results of 1978, 1981 magnetic orientation experiments would possibly be due to magnetic switching artifacts, just because Johnsen and Lohmann (2005, 2008) said so is clear proof that Newton & Kajiura did not carefully study my original articles on the topic, Kalmijn (1978b, 1981).

Hence, that sharks can be trained to respond to a magnet hidden in the bottom sand of the observation tank, is no evidence of a magnetic sense, as bar magnets are not common sources of magnetic fields in the oceans' natural habitat and it is not clear why the animals should attack them unless they *mistake* them for something else, predator, prey, or conspecific for instance. It only proves that animals moving in a magnetic field induce electric fields and that nonuniform localized electric fields are readily mistaken by hungry sharks and rays for potential prey. In the first five sessions depicted in their Fig. 1.2, about one-third of the animals responded to the hidden magnetics even before being successfully trained to do so. Hence, the responses to bar magnets are mere *biological artifacts*. The responses are consistent with an electric sense and with a magnetic sense but are no proof of either one. Fortunately, such hidden magnets are uncommon in the habitat of sharks and rays, unless the scientists for their entertainment put them in there to fool the animals.

Newton and Kajiura's (2020) reinvestigated the question whether stingrays may detect the polarity of magnetic fields. Sure, Kalmijn (1978b) had proven stingrays capable of orienting in magnetic fields of natural strength, proving that they not only detect the direction, but also the polarity of the fields.

However, to quote Newton & Kajiura: "Kirschvink (1989) argued that the presence of inadvertent chemical, visual, and tactile cues and magnetic gradients were potentially confounding variables that might better explain the results of Kalmijn (1978b)." Independent of the fact that Kirschvink's remarks were incorrect and not credible (see elsewhere), they explicitly were not meant to throw any doubt on the magnetic orientation of the stingrays, to the contrary, but only on whether the animals relied on the motional electric principle or, perhaps, on some other unknown magnetic sensory system. Hence, the reasoning of Newton & Kajiura demonstrates a lack of carefully reading the original literature. Anyhow, Newton & Kajiura's experiments used highly nonuniform magnetic test fields, biologically invalidating their inferences concerning orienting in the earth's magnetic field. Moreover, the applied field added to the earth's magnetic field resulted in field lines [they are not isolines] bending either to the right or left on approaching the horizontal part of the T, depending on the polarity of the field produced by the coil system. Hence, independent of the polarity of the field, the animal would simply have to follow the field lines to yield the same results as those by Newton & Kajiura's accredited to the polarity of the field.

4.9 The Literature Steadily Slides Even Steeper Downhill

Passive versus active electro-orientation

Newton et al. (2019) are apparently not aware of the fact that sharks operate their electric sense in the *active mode* when orienting in the earth's magnetic field, themselves motion-electrically generating the energy of the electric fields they detect, and, conversely, weakly electric fishes in prey detection operate their low-frequency electric sense in the *passive mode*, the prey bioelectrically generating the energy of the fields the sharks detect, even though Kalmijn has explained this on the first page of his 1974 handbook chapter. He did not invent the conventional terms active and passive but learned them from the established sensor-engineering literature. That other biologists have used the terms active and passive incorrectly is no excuse for mixing up the issue even more. Note also that it is the sensory system, including the behavioral feedback, that operates in the active or passive mode. The ampullae of Lorenzini *per se* are in a certain sense active, as they detect the signals at the cost of their own energy. Actually, they do not draw a positive signal current but actively produce a negative, signal-enhancing current in the opposite direction driven by positive feedback.

The true story of electrical sensitivity

In order to judge whether the prevailing oceanic electric fields are detectable by sharks and rays, it is of the greatest importance to know how sensitive the animals really are. Kalmijn (1982) showed the stingray *Urolophus halleri* to be perfectly capable of orienting in uniform electric fields at voltage gradients of 5 nV/cm, but not lower. In detecting prey-simulating dipole fields, the shark *Mustelus canis* responded in night-time experiments conducted in the wild at sea even from distances at which the voltage gradients were as low as 1–2 nV/cm, but the majority of the responses were from distances at which the voltage gradients were 4 nV/cm or higher. Reports of responses to electric fields of less than 1 nV/cm are based upon the use of an erroneous calibration formula, as I will explain in detail because of the crucial importance of reliably knowing the lowest voltage gradients at which sharks and rays can operate to avoid wild speculations rendering the shark's electric sense incredible.

As to the behavioral calibration of the electrical sensitivity of sharks and rays, according to Kajiura (2001) and Kajiura and Holland (2002), the electric field strength of a dipole would not only decrease as the cube of the distance from the source but also vary as a cosine function with respect to the dipole axis, so that the field strength would be greatest along the dipole axis and decrease to zero when reaching the normal symmetry plane, WHICH IS NOT TRUE AT ALL. The radial component decreases to null at 90 degrees, but the tangential component reaches at 90 degrees half the value of the axial component along the dipole axis, while pointing in the opposite direction. Hence the electric field strength decreases only by a factor of two from an angle of 0 to 90 degrees with the dipole axis, as explicitly stated by Kalmijn (1982), whose dipole formula Kajiura and Holland (2002) corrupted. In particular, they mistake the radial directional derivative of the voltage for the voltage gradient. Of course, the radial component is zero in the central equipotential plane, but the tangential component is normal to the central equipotential plane and is not zero at all, no matter whatever physics textbook they consulted.

The disastrous consequence is that—up to the very present—all sensitivity data reported by Kajiura and his followers, based upon his erroneous dipole calibration formula, are wrong, and so are all scientists that have quoted the invalid sensitivity numbers without checking on their validity. No wonder that his coworkers can get sensitivities as high as they wish using Kajiura's formula. Just wait for an approach within the central equipotential plane, which according to Kajiura would mean an infinitive sensitivity. "Elementary" Sherlock would say. Kajiura actually blamed me by saying that he learned the dipole calibration formula from me but he bungled it by using only the radial component, apparently not aware of the tangential component. Even his doctoral thesis committee chair, Dr. Holland, let it pass. Sharks know better, though, at least they behave as if. Hence, the only reliable sensitivity data are those collected by Kalmijn (1982) over a period of 30 years: for DC uniform ambient electric fields 5 nV/cm and for DC prey simulating dipole fields as low as—but not lower than—1–2 nV/cm unless there are hard sensitivity data that I am not aware of.

Scholium

A physical electric dipole field is defined by the dipole moment **p**, a vector in the direction from the – pole to the + pole, of strength *Id*, the dipole current *I* multiplied by the distance *d* between the – pole and the + pole. In the limit of *d* tending to zero, at constant **p**, and at uniform resistivity ρ, the physical dipole field becomes a mathematical dipole field, of which the scalar potential Φ at a point *P* is given by:

$$\Phi = \frac{\rho(\mathbf{p}\cdot\hat{\mathbf{r}})}{4\pi r^2} \quad \text{or} \quad \Phi = \frac{\rho Id\cos\theta}{4\pi r^2}$$

with $(\mathbf{p}\cdot\hat{\mathbf{r}})$ the scalar product of dipole moment **p**, the unit radius vector $\hat{\mathbf{r}}$ pointing from the center of the dipole to the point *P*, and the cosine of the angle θ between the direction of the radius vector and the dipole moment, and *r* is the distance to the point *P*.

The negative of the voltage gradient $-\nabla\Phi$ gives the oppositely directed field strength **E** of the dipole field, expressed in the radial and tangential components of spherical coordinates:

$$\mathbf{E} = \frac{\rho p\cos\theta}{2\pi r^3}\hat{\mathbf{r}} + \frac{\rho p\sin\theta}{4\pi r^3}\hat{\boldsymbol{\theta}}$$

Note that in half space, the field is twice as strong at the same dipole moment, and that along the dipole axis ($\cos\theta = 1$, $\sin\theta = 0$), the field is twice as strong as the field in the normal plane ($\cos\theta = 0$, $\sin\theta = 1$).

The strength of the field at a point P is:

$$E = \sqrt{\left(\frac{\rho p \cos\theta}{2\pi r^3}\right)^2 + \left(\frac{\rho p \sin\theta}{4\pi r^3}\right)^2}$$

Vectorially adding a uniform field $\mathbf{E}_a$ to the dipole field $\mathbf{E}_d$ gives a total field $\mathbf{E} = \mathbf{E}_a + \mathbf{E}_d$, which also describes the field outside an insulating sphere in a uniform field.

What went wrong with Kajiura's dipole formula?

Since Kajiura's erroneous electrical sensitivity figures have been widely quoted, the reader may wonder how it is possible that in hindsight he used the wrong dipole calibration formula. Kajiura even mentioned that he was the first one to report that he had actually verified the dipole field the animal detects by direct measurement. Too bad, that is where he went wrong. Except that he made the measurements, not in half space but in only a small volume of seawater, he measured the voltage distribution of the dipole field, not the voltage-gradient distribution the animal detects. Since the voltage is a cosine function of the angle between the dipole axis and the radius vector of the point of observation, he felt urged to add a cosine function to the voltage-gradient formula as well. As a result, when the shark would approach the dipole source in the normal plane, the field it detects would be zero, rather than half the value of the field along the dipole axis, as explicitly stated in Kalmijn (1982). From the above Scholium, it is clear that Kajiura (2001) simply "overlooked" the tangential component of the dipole field. By studying Kalmijn (1988b, 1997, 2000) approach algorithm, Kajiura should have noticed that something in his own formula had to be wrong.

Do-not change history

True, the electrical sensitivity of the ampullae of Lorenzini was first noted by Murray (1960a, b). Yet, he started his subsequent 1962 full-length article saying, "The biological function of the ampullae of Lorenzini of elasmobranch fish remains uncertain," as they respond to temperature changes, mechanical stimuli, salinity changes, and now also to electrical stimuli. As to the function of the ampullae of Lorenzini, Murray ironically said: "Among the possibilities are the detection of the weak currents which result from the movement of conductors through the earth's magnetic field, although it would appear to be about one order of magnitude too small for the ampullae to detect under normal conditions." True, Dijkgraaf and Kalmijn (1962) reported the sharks' behavioral responses to electric fields without suggesting any biological significance, as I preferred to report the hard facts, but not to speculate. The breakthrough came when Kalmijn (1966, 1971) measured the DC and low-frequency bioelectric fields of aquatic animals and proved behaviorally that sharks and rays zero in on the source of the fields in predation. The bioelectric detection of prey constituted the first proof that sharks and rays have a genuine electric sense, for which Murray personally complimented the present author. The introduction of the electric and magnetic theory of motion-electric orientation in sharks and rays followed a few years later, Kalmijn (1974). If you wish to review the literature do so with greater care.

On and on

As for the theory of motional electric orientation in the earth's magnetic field, Newton, Gill, and Kajiura mention Paulin (1995) on the same line as Kalmijn (1974) without warning the reader that the two theories are mutually exclusive, as each of the two originators considers the alternative theory fatally flawed. Hence, Newton, Gill, and Kajiura implicitly admit that they have not studied or do not understand either theory, hence are not able to distinguish the chaff from the wheat. Concerning the operation of the ampullae of Lorenzini as it is utter nonsense as to how the electroreceptor epithelia remotely sense the charges at the skin. The ampullae of Lorenzini act as ideal voltmeters, not detecting charges but voltages, regardless of how they are produced. For that, Newton, Gill, and Kajiura should have consulted the original articles on the topic.

To quote Newton, Gill, and Kajiura: "Kalmijn (1982) and Pals et al. (1982) demonstrated that some species of elasmobranchs can be behaviourally conditioned to orient toward electric dipoles during the onset of a DC field and can distinguish electrical gradients of c. 5 nV cm^{-1}." All wrong again: Kalmijn studied spontaneous responses to DC dipole fields, whereas Pals et al. did not report a sensitivity of 5 nV/cm. Ever since his original publication of 1971, Kalmijn has observed many kinds of sharks and rays in the laboratory and at sea to orient *spontaneously* in electric dipole fields simulating the bioelectric fields of prey, that is, without training, and observed them to do so even when the field was purely DC, not just at the onset of the field, but all the time. Thus, in night-time experiments at sea on the shark *Mustelus canis*, Kalmijn (1982) determined threshold gradients as low as 1–2 nV/cm. The 5 nV/cm threshold comes from training experiments in *Urolophus halleri* in uniform electric fields faithfully simulating the motionally induced fields of ocean streams, Kalmijn (1978b, 1981). Another erroneous quote from Newton, Gill, and Kajiura: "The range of the induced electric fields produced by the physical movement of conductive seawater (c. 500–8000 nV/cm) through the Earth's geomagnetic field," which is two, respectively, one order of magnitude off.

Concerning the electrical detection of ocean streams: "Chondrichthyans could use this method to passively determine their orientation within oceanic and tidal currents (Paulin 1995)." That is what Kalmijn (1974) proposed based on hard physical and biological evidence, whereas in traveling sharks Paulin downplays the ambient-electric field information. "Additionally," referring to Kalmijn (2000) and Paulin (1995) concerning the detection of the earth's magnetic field "it has been hypothesized that the Ampullae of Lorenzini might detect location and directional cues from the GMF and possibly use them to actively orient and navigate during migrations." Newton, Gill, and Kajiura should tell us which author to believe and why. One more astonishing quote, "However, if chondrichthyans use their electroreceptors indirectly to detect magnetic fields, then it is unclear [at least to Newton, Gill, and Kajiura] how they might distinguish between magnetic and electric cues." Did not Kalmijn over and over emphasize that the electrical signals received from the ambient electric fields and the electric fields the animals themselves motionally induce are entirely different functions of their exploratory movements and did he not discuss those differences in great detail?

As to their Fig. 2.6: (1) Bioelectric fields of prey have not and cannot have the configuration of a DC monopole field as shown in the picture by the equipotential surfaces. (2) Motionally induced electric fields due to rectilinear motion leave and re-enter the moving animal, rather than completely staying inside. (3) Earth's magnetic field lines are at the scale of the figure certainly not curved noticeably, unless there are large magnetic bodies nearby. (4) Electric field induced by AC magnetic field of cable depicted overly simplistic, hence misleading. In conclusion, the whole picture reveals that the authors either do not know the configurations of the electric and magnetic fields or are unable to picture them properly. I hope that the wind-farm engineers know better, at least about the wind farms. The sharks are not their expertise. Whose is it? They claim they want to protect sharks from the electric and magnetic fields wind farms produce, not aware of the most offending kind of electric fields they still have not yet discovered, and all wind farms suffer from.

As to the polluting effects of undersea powerline and windmill farms, Newton, Gill, and Kajiura apparently have not yet recognized the most offending kind of electric fields all wind farms suffer from, despite my repeated warnings. True, I have not yet told them what to look for.

Once more, conditioned behavioral responses to biologically inadequate electric fields induced by a change in the magnetic field or by motion in any other magnetic field than the earth's field or steady faithful simulations thereof are not evidence for a biologically genuine magnetic sense. They are artifacts, just as the threshold sensitivity values obtained based on a corrupted calibration formula are artifacts. We may learn something from those artifacts, but not that the animals have a magnetic sense. I hope that I have made this clear enough to have it sink in once and forever. I tried my hardest.

Not too fast though

At this stage, the author feels so disheartened about the continuous stream of ever-worsening misinformation that for the moment he gives up patrolling the pertinent literature. His writings certainly are not the ultimate truths either, but merely serious attempts to get it all straight. He dutifully blew the whistle. Nevertheless, please, let me know where I went wrong so that I can correct my mistakes for the sake of science and in deference to my dear animals. Was the author's lifetime effort in vain or may the present article save science from further perpetuating these gross errors?

Newton and Kajiura (2020)

Having learned of the bi-coordinate grid of the earth's magnetic field from Gould (1998), Newton and Kajiura (2020) wonder whether stingrays may use the magnetic intensity and the magnetic inclination of the earth's magnetic field as two intersecting coordinates systems to orient on the earth globe, a possibility critically discussed by Phillips (1996). Hence, they raised the fundamental question whether stingrays can appreciate differences in magnetic intensity as distinct from differences in magnetic inclination. I agree, the principle of motional electric detection of the earth's magnetic field by sharks in ocean waters leaves no doubt about its *theoretical* possibility, but is it really feasible?

To test the *resting* stingray's sensitivity to differences in magnetic intensity, the authors decide to apply static magnetic fields differing by a factor two in strength in quick succession to simulate the motion of the animal in a magnetic field, a reasoning that escapes me altogether. To get the stimulus in the frequency range of the animal's electrical sensitivity, they switch every second between the magnetic intensities of parallel fields. To test the stingray's sensitivity to differences in inclination, they follow the same procedure, now switching every second between two equally strong magnetic fields differing in inclination by 60 degrees.

What are the motionally induced electric fields the animal detects in the two different situations? To switch back and forth between the normal and the double-strength magnetic field, they apparently add once a second for a second a parallel magnetic field to the undisturbed earth's magnetic field. To switch back and forth between the normal and the differently inclined magnetic field, they apparently add once a second for a second a magnetic field under an angle with the undisturbed earth's magnetic field. Hence, in both cases, they add once a second for a second a magnetic field of different strength, either in the direction of the earth's magnetic field or under an angle with the earth's magnetic field.

What will be the effect of the magnetic fields added to the undisturbed earth's magnetic field? Well at the onset of the applied magnetic field, there will be a magnetic transient in the observation tank, causing a volume electric current eddy around the axis of the added magnetic field, and when the applied magnetic field is switched off again, there will be a similar magnetic transient in the observation tank, causing similar transient volume electric current eddy but in opposite direction. The eddy currents will result in electric fields that the animal may detect, but in no way simulate an electric field the animal would detect when moving in the presence of a static magnetic field. On the contrary, when the animal detects the magnetic field by moving itself, it produces in our frame of reference a semi-dipolar local electric field of which the animal itself is the source.

In brief, the article is based on physically inadequate stimuli, that is, on artifacts, which do not support the conclusions. Having said this, I can only ask Newton and Kajiura to stop teasing the animals with seaquakes of different but biologically unnatural non-sensical configurations, not in any logical manner reproducing the electrical stimuli received from topographic magnetic intensity and inclination changes of the earth's magnetic field that the authors believe they represent. This may not come as a surprise, as all articles of the authors on the magnetic sense of sharks and rays are based upon a total lack of proficiency in the governing physics, and their experiments are all based upon biological artifacts in the strictest sense of the word, as I have reiterated repeatedly. I am sorry to say so. I wish, I were wrong, but unfortunately, I am not.

Finally, some good news! Keller et al. 2021 reproduced one of my very first observations on the magnetic sense of sharks, thereby confirming the results. Well, "reproduce" is not the right term, as they do not mention my original observations, even though they list the article in which they were described. After my initial circumstantial findings, I continued conducting the very first experiments definitely proving that sharks and rays have a true magnetic sense. Keller *cs* went further and speculated far beyond what they could justify on the basis of their observations. Well, a little speculation does not hurt, as long as the readers are aware that speculation is speculation, not hard science.

Both my and their experiments showed that, when sharks are magnetically displaced south over a few hundred miles, they tend to go north, back where they came from. Although experimentally the results were essentially the same, they show interesting differences, rendering them complimentary.

1. My leopard sharks were physically moved over a distance of a few hundred miles from a place north to a place south, featuring significantly different magnetic strengths and inclinations, whereas his bonnethead sharks were magnetically moved by changing the ambient magnetic field in their home tank to the magnetic field prevailing a few hundred miles south.
2. My leopard sharks tended persistently to move north, even when held in differently sized observation tanks, 6 and 20 ft in diameter, and when the largest tank was physically rotated, hence showing very robust results, whereas their bonnethead sharks significantly tended to move north after changing the magnetic field in one and the same home tank.

On a more sour note, their comments on the literature makes one wonder. The possibility of sharks using their electric sense "to glean information from the earth's magnetic field" as they say was not brought up in Kalmijn in 1971, the article in which I proved sharks and rays to have an electric sense, but in Kalmijn (1974), where I introduced the theory of motional electric orientation in the earth's magnetic field. To refer to Paulin without the slightest reservations proves that the authors do not understand either one, the original rigorous version and the more recent version, otherwise they should have made a qualifying remark as to who to believe, either the one or the other. Concerning "the elasmobranchs' ability to discriminate between different components of the earth' s magnetic field" and the statement that sharks and rays "have been trained to respond to geomagnetic polarity and intensity," they properly quote me, but again the wrong article, Kalmijn (1982) instead of Kalmijn (1978b). However, they furthermore refer to the several articles published by Holland, Anderson, Newton, and Kajiura, all of whom have—by my definition—merely studied biological artifacts, to wit, the behavioral responses of sharks and rays to the electric fields induced by moving in the highly *nonuniform* fields of permanent magnets, rather than motionally interacting with the locally highly *uniform* earth's magnetic field or faithful simulations thereof. The responses to permanent magnets or to small coils are merely consequences of the animals having an electric sense, mistaking the magnets for prey or other local objects, whereas the responses to uniform magnetic fields give the animals a sense of direction in space for open-ocean orientation.

4.10 Concluding Remarks by the Author

A confession

The first section of the present article consists of a scholarly account of my endeavor to explain to myself what I originally advanced in 1974 concerning the motional-electric orientation of sharks and rays in the oceans' ambient electric and magnetic fields, long before I knew that the animals indeed practice in daily life what I envisioned on the basis of Faraday's motion-electric principle and the astounding electrical sensitivity of the animals I had just established. Hence, the first section is not a tutorial, but a string of thoughts that went through my searching mind, put in writing in the hope that the reader may benefit from it, at least if he or she is not turned off by the winding discourse. My initial articles may be more easy reading. Although more succinct, they are still equally valid, except that here and there I might not have been emphatic enough about possible misunderstandings, which I later became aware of, after learning what did and did not come across properly.

The second section seeks how sharks and rays behaviorally derive the sought-for information from the biologically significant features of the electric fields they receive. That is, what exploratory moves they may make in support of their electric and magnetic sense. It is meant to prove that it is possible for the animals to sense their drift in ocean currents and to sense their magnetic compass headings side by side by executing the elementary individual moves envisioned by me. Yet, it is understood that the animals may seamlessly string together the individual moves during their normal behavior of locally wandering about or traveling long distances. Hence, my account suggests how sharks and rays may motionally behave to receive the desired sensory information but leaves it to the animals to decide how. They certainly orient electrically and magnetically extremely well. For them, it evidently is less difficult to practice the intricate physics than for us to preach it.

The third section is an embarrassing shame. I was taught physics and biology in the rigorous, old-fashioned academic fashion at the prestigious University of Utrecht, The Netherlands. I have kept to my principles, but by doing so I have lost confidence as to how science in my particular field of research is conducted nowadays. Hence, I faced the tough question, shall I speak up or just keep quiet. The latter would cramp my style. It is unlikely that I will change the confused minds of my colleagues, by now old-timers as well. Yet, one may fight for or against the truth but one cannot change it, however hard one tries. My ultimate goal is to reach the younger generation of inquisitive students. I remain committed to fulfil my holy duty in deference to those who have supported my scientific effort, both academically and financially.

Do sharks and rays have an electric sense?

Is it justified to state that sharks and rays have an electric sense? "Yes" we may say, but only if we make the tacit reservation that, perhaps, some of them do not. There are no reports of sharks and rays that lack ampullae of Lorenzini altogether, but the manta ray certainly has a greatly reduced number of them. Is it justified to state that sharks and rays orient in the oceans' ambient electric fields and in the earth's magnetic field? Here, we have to be more careful. That they definitely do so has been rigorously proven only for the round stingray, *Urolophus halleri*. Even in this case, the question has been raised whether it is statistically justified to infer that all round stingrays can do so, if it has only been proven to be true for a limited, arbitrarily collected number of them. However, the latter reservation goes too far for me. That such astounding sensory abilities would only be present in a few specimens of a certain species seems preposterous to me. Indeed, there are numerous, though often more casual observations of sharks and rays orienting in ambient electric and magnetic fields. True, even if all sharks and rays are equal in this respect, they need not all be equally equal. Depending on the natural history of the species, the one may rely on its electric sense more for predation, and the other more for open-ocean orientation, but the two are certainly not mutually exclusive.

What about a motion-electric magnetic sense?
Is it justified to say that sharks and rays orient magnetically by use of their electric sense? That is a legitimate question, though it is a well-established fact that, by moving in the earth's magnetic field, sharks and rays receive electrical signals well within their sensitivity and frequency range. These fields are for sharks and rays of the same kind as the ambient electric fields we know for sure they orient in, except that they are very different functions of the animals' exploratory movements. After theoretically proving that sharks and rays might well be able to orient magnetically by the use of motional electric fields, yes indeed, the round stingray proved to be able meaningfully to orient in the earth's magnetic field. Whether the animals might have ignored the clear motional electric signals informing them of their magnetic compass heading in favor of an alternative signal that we do not know of, I leave that for the nay-sayer to ascertain. Yet, I assume that sharks and rays rely on their electric sense in orienting magnetically until it is proven that they do not. To verify that they do, I have initiated behavioral experiments in which stingrays are conditioned to orient in a magnetic null field in response to the strength and polarity of a uniform vertical electric field simulating the electric fields the animals would have received if the ambient magnetic field were still present. The results will be reported as soon as I have the opportunity to complete the contemplated experiments.

Certainly, there are several misleading experiments by eager authors that bend over backward trying to disqualify my 1978 behavioral experiments striving to become the ultimate authorities of the present time. In my opinion, for an animal to be considered to have a magnetic sense, one has to prove it to make biologically meaningful use of naturally occurring magnetic fields. In the instance of sharks and rays, that means making biologically significant orientational use of the earth's magnetic field or faithful simulations thereof, independent of how the earth's magnetic field is detected. I consider any responses to magnetic fields other than the earth's magnetic field or faithful simulations thereof biological artifacts. True, biological artifacts are facts, but not the kind of facts we need to prove sharks to have a magnetic sense. At best, we may argue it to be circumstantial evidence, but the critical step is to advance from circumstantial to hard evidence.

First phase of mission to be accomplished. What is next?
Having spent many hours on preparing the present manuscript, I have covered only a small part of what I would like to tell you, not only about the electric sense of sharks and rays, but also about the lateral-line and inner-ear sense organs of fishes in general. The system of the electric sense and the system of the inner-ear organs have a lot in common from a physical as well as a physiological point of view. In particular, the bioelectric fields and the water-motion fields of the sharks' prey are similar multipole fields that in predation can be relied on by the same approach algorithm Kalmijn et al. (2002). However, the very receptor cells are structurally rather different, as they are adapted to the detection of stimuli of different physical nature, electric and magnetic versus mechanical, that is, hydrodynamically and low frequency inertially. Yet, the two sensory systems are operating side by side, both peripherally and centrally. Hence, they are expected to provide mutual support, complementing one another, in addition to being aided by smell and vision. My effort has been to identify the physical stimuli and the behavioral responses. As to the central nervous system, yes, the relationship between the two is a matter of brain processing. Hence, I know the sensory task of the central nervous system, but I certainly do not know how it processes its task. I gladly leave that to my more cerebral colleagues.

4.11 Concluding Remarks by Sharks and Rays

Let us ask the animals their points of view
"The ampullae of Lorenzini enable us, marine sharks and rays, to sense and respond to DC and low-frequency electric fields, commonly at a mere 5 nV/cm, at pressing times even at 1–2 nV/cm. In our ocean habitat, we are not only subject to the local electric fields of objects that produce electric fields, such as prey, or distort ambient electric fields, such as rocks, but also to uniform electric fields of two kinds, both of which are well within our sensitivity range but differ profoundly in the directional information they offer. Though the two kinds of fields are both uniform, they are radically dissimilar in how they appear to us in response to our exploratory movements. The one kind informs us, among others, about the upstream direction and strength of ocean streams, while the other kind informs us of, what you call, our magnetic compass headings. Certainly, to use either one or both kinds of fields for open-ocean orientation, we must be able to tell them apart, which we certainly can based on the physics we practiced long before it was recognized by a man Faraday and explained by a man Einstein. We have acknowledged this in the behavioral sessions with our human friend from The Netherlands, in which we proudly proved our ability to establish our bearings in ambient electric as well as in ambient magnetic fields faithfully simulating those prevailing in open-ocean waters. Your oceanographers may mimic us, for instance, by turning our friend's Boston Whaler left or right at constant forward speed to detect the one kind of field, your scientists call electric, and by accelerating rectilinearly to detect the other kind, your scientists call the magnetic."

"Even though we, marine sharks and rays, pride ourselves on our keen electric sense, there are some of you, so-called biologists, who believe that we are still not sensitive enough to detect electric fields less than 20 nV/cm. That means, we would miss out on a lot of vital information. Why did these offending naysayers so pitifully underrate our electrical sensory abilities? Did they ask us the wrong questions or query us under such stressful experimental circumstances that we were incapable of performing? At the other extreme, there are the overly gullible believers, actually a whole bunch of them, who keenly observe us initiate our electrically guided feeding attacks at measured distances from prey simulating dipole sources, but report entirely different field strengths than we experience, often even lower than we are able to detect. Our human friend explained to us that the latter authors use an erroneous calibration formula, confusing the angular dependences of voltages and voltage gradients in dipole fields. Don't they know? We are proud to be as sensitive as we are, but we do not pretend to be even more sensitive. Admittedly, we do not understand your mathematical physics, but we already knew how to apply it long before your scientific forefathers formulated it only a few hundred years ago. Perhaps, we can still teach scientists a lesson or two."

"An even more troublesome quandary has arisen. Our human friend, who originally proved us to have a genuine electric sense and the ampullae of Lorenzini to be our electroreceptors, and who always sought the advice from his physics forefathers and from us, sharks and rays, is now himself under attack. A weird, often recited adversary by the name Paulin 'proves' him all wrong and offered a "totally new" alternative physics foundation and a rehashed motional electric orientation theory that must have made Einstein and his predecessors turn over in their graves. The naïve admirers of the adversary are pleased to hear that they are not the only ones who do not understand our unique electric and magnetic sensory abilities. Whoever these critics may be, we sharks and rays certainly do not agree with the reasoning of this demeaning critic, neither will we will follow his senseless method of orienting in, what you call, the earth's magnetic field. We would not know how to, as he tells us to discard the most significant electrical stimuli we receive and to dig up the remnants from under the noise to get the humpty-dumpty signal together again. We are not that foolish."

"You may wonder why we detect not only the electric fields of ocean currents, as if that is not difficult enough, but also the motional electric fields we introduce ourselves by our own motional activity. One might think that such dual use makes it extra complicated for us. True, but we cannot do otherwise, since, as we told you already, for us the two kinds of fields are of the same configuration and of similar strengths, as your Einstein taught you. If we detect the one kind, then we cannot escape detecting the other, and *vice versa*. As a consequence, our central nervous system receives vital information from both types of fields. Nevertheless, though nobody objects to us orienting in the one kind of electric fields, those induced by ocean currents, your scientists squabble whether, when we determine our compass headings, we really rely on the other kind of electric field, the one we induce ourselves by our exploratory movements. Sure, we could opt to ignore the compass information the latter offers and rely on another sense organ providing the same information, based on another physical principal, the kind you call magnetic. But why would we? That would make life really complicated. True, we may have an *additional*, non-induction magnetic sense to orient for instance to detect nonuniformities in the magnetic field. That is for your scientists to find out."

"Even more unbelievable, one of your scientists, not comfortable with one single sensory system detecting two kinds of uniform electric fields, went so far as to argue that sharks and rays *must have* an independent, non-induction-based magnetic sense enabling them to reject the confusing electric signals the induction-based mechanism produces when we orient in ambient electric fields or in the fields of prey. Those anti-induction-based theories, if correct, would show us to be more sophisticated than our soul mate ever imagined. Well, that is the grand idea some of your scientists brought up, based on the results of putting magnets in the nares of stingrays. Indeed, it would be a surprise if we would make great effort to correct for the fields due to our motion in the earth's magnetic field, instead of appreciating them for the compass information they provide. It is no great surprise, however, that those magnet scientists are dead wrong, because they do not know their physics as well as we practice it day and night. Hence, from their misinterpreted magnetic interference experiments we have little to fear of being denied our induction-based magnetic sense. To the contrary, these magnet scientists unwittingly provided the strongest evidence in favor of the motional electric principle."

"Recently, it has gotten even worse: another series of cruel conditioning experiments performed on us, stingrays. To test whether we are able to use a bi-coordinate system based on the local magnetic intensity *versus* the local magnetic inclination to determine our location on the earth's globe, they harassed us with two kinds of electrical seaquakes totally unrelated to how we motional electrically detect the earth's magnetic field. Hard to believe, but they compared our behavioral reactions to changing the intensity of the ambient magnetic field every second by a factor of two *versus* switching the ambient magnetic field every second back and forth over an angle of 60 degrees. However, they did not realize that physically they instead induced overly strong transient electric fields in the whole test basin that had nothing to do with the magnetic field changes we would according to them detect. Perhaps, the present article of our friend can help you, the reader. Anyhow, he consulted with us by means of behavioral experiments for more than half a century, asking us the right electric and magnetic questions that we behaviorally answered to the best of our abilities. Admittedly, he has a sharp tongue, but he has to speak loudly and clearly to be heard above the noise of the utterances made by his fast-talking adversaries, who keep offending us. He endeavors to show the beauty of the beasts we are. We gladly join him in teaching our common foes a lesson or two."

"Luckily enough, one team of shark enthusiasts repeated an old experiment of our friend, proving that we sharks tend to return home after being displaced magnetically. The only catch is that they did not realize—at least did not acknowledge—that long ago essentially the same observations had been made by our friend—who discovered our electric and magnetic sense in the first place—even though he included his observations in the same article in which he published his classical experiments establishing our magnetic sense unambiguously once and forever."

"In conclusion, your scientists may consider us dumb and rude. Admittedly, in some respects, though not all, humans seem smarter than we are, but we seriously doubt whether they are more civilized. Our friend agrees. Let us see who will last the longest, humans, sharks, or cockroaches."

References

Akoev GN, Ilyinsky OB, Zadan PM (1976) Responses of electroreceptors (Ampullae of Lorenzini) of skates to electric and magnetic fields. J Comp Physiol 106:127–136

Anderson JM, Clegg TM, Veras LVMQ, Holland KN (2017) Insight into shark magnetic field perception from empirical observations. Scientific Reports

Andrianov GN, Brown HR, Ilyinsky OB (1974) Responses of central neurons to electrical and magnetic stimuli of the Ampullae of Lorenzini in the black sea skate. J Comp Physiol 93:287–299

Bezrukov SM (2003) Sensing Nature's electric fields II: ion channels as active elements of linear amplification. In: Bezrukov SM (ed) Unsolved problems of noise and fluctuations, vol 665. American Institute of Physics, pp 142–149

Blakemore R (1975) Magnetotactic bacteria. Science, New Series 190(4212):377–379

Bodznick DB, Montgomery JC (2005) The physiology of low-frequency electrosensory systems. In: Electroreception. Springer, pp 132–153

Bodznick DB, Montgomery JC, Tricas TC (2003) Electroreception: extracting behaviorally important signals from noise. In: Sensory Processing in Aquatic Environments, pp 389–403

Brown HR, Ilyinsky OB (1978) The ampullae of Lorenzini in the magnetic field. J Comp Physiol 126:333–341

Dijkgraaf S, Kalmijn AJ (1962) Verhaltungsversuche zur Funktion der Lorenzinischen Ampullen. Naturwissenshaften 17:400

Dijkgraaf S, Kalmijn AJ (1963) Untersuchungen über die Funktion der Lorenzinischen Ampullen an Haifischen. Z Vergl Physiol. 47:438–56. True to life drawings (K) of ampullary and lateral-line system in *Scyliorhinus canicula* and proof (K & D) of ampullae serving as electroreceptors

Dijkgraaf S, Kalmijn AJ (1966) Versuche zur biologischen Bedeutung der Lorenzinischen Ampullen bei den Elasmobranchiern. Z vergl Physiol 53:187–194

Dixon WJ, Massey FJ (1969) Introduction to statistical analysis, 3rd edn. McGraw-Hill, pp 362–374. Chapter 18, Sequential Analysis

Einstein A (1905) Zur Elektrodynamik bewegter Körper. Annalen der Physik 17:891–921. See also: On the electrodynamics of moving bodies. In: The Principle of Relativity. Translated by W. Perrett and G. B. Jeffery, 1952, Dover edition, pp. 35–65

Faraday M (1832) The Bakerian Lecture: experimental researches in electricity. Second Series. Phil Trans R Soc Lond 122:163–194. See also: 1965 Dover edition of Experimental Researches in Electricity. Vol. I, pp. 42–75

Faraday M (1839) Experimental Researches in Electricity. Fifteenth Series. Philosophical Transactions of the Royal Society of London, Vol. 129, pp. 1–12. See also: 1965 Dover edition of Experimental Researches in Electricity. Vol. II, pp. 1–17. *Electric force of Gymnotus*

Gould JL (1998) Sensory bases of navigation. Curr Biol 8:R731–R738

Hammond P (1985) Electromagnetism for engineers, 3rd edn. Pergamon Press, Oxford

Johnsen S, Lohmann KJ (2005) The physics and neurobiology of magnetoreception. Nat Rev Neurosci 6:703–712

Johnsen S, Lohmann KJ (2008) Magneto reception in animals. Phys Today 2008:29–35. https://doi.org/10.1063/1.2897947

Johnson CS, Scronce BL, McManus MW (1984) Detection of DC electric dipoles in background fields by the nurse shark. J Comp Physiol A 155:681–687

Jungerman RL, Rosenblum B (1980) Magnetic induction for the sensing of magnetic fields by animals—an analysis. J Theor Biol 87:25–32

Kajiura SM (2001) Electroreception in Carcharhinid and Sphyrnid sharks. Ph.D. Dissertation, University of Hawai'i, p 110

Kajiura SM, Holland KN (2002) Electroreception in juvenile scalloped hammerhead and sandbar sharks. J Exp Biol 205:3609–3621
Kalmijn AJ (1966) Electro-perception in sharks and rays. Nature 212:1232–1233. Transient cardiac decelerations of *Raja clavata* at 0.01 μV/cm. Ventilatory potentials of flounder *Pleuronectes platessa*. Electrically elicited feeding responses in *Scyliorhinus canicula* and *Raja clavata*
Kalmijn AJ (1971) The electric sense of sharks and rays. J Exp Biol 55:371–383. Bioelectric fields of prey evoke well-guided feeding attacks, proving *Scyliorhinus canicula* and *Raja clavata* to have a genuine electric sense. First suggestion of electrical detection of drift in ocean streams
Kalmijn AJ (1974) The detection of electric fields from inanimate and animate sources other than electric organs. In: Fessard A (ed) Handbook of sensory physiology, vol III/3. Springer, Berlin, pp 147–200. Review of earlier research and introduction of his theory of motional electric and magnetic orientation in sharks and rays. Behavioral frequency response of the shark *Scyliorhinus canicula* to dipole fields at frequencies from DC up to 4 Hz
Kalmijn AJ (1977) The electric and magnetic sense of sharks, skates, and rays. Oceanus 20:45–52. A nontechnical narrative concerning the research conducted on the function of the elasmobranchs' electric and magnetic sense in prey detection and open-ocean orientation
Kalmijn AJ (1978a) Electric and magnetic sensory world of sharks, skates, and rays. In: Hodgson ES, Mathewson RF (eds) Sensory biology of sharks, skates, and rays. Government Printing Office, Washington DC, pp 507–528. Electrically evoked feeding responses of *Mustelus canis* in the wild and behavioral proof of magnetic orientation in *Urolophus halleri*
Kalmijn AJ (1978b) Experimental evidence of geomagnetic orientation in elasmobranch fishes. In: Schmidt-Koenig K, Keeton WT (eds) Animal migration, navigation, and homing. Springer, New York, pp 347–353. Magnetic orientation experiments, theory and evidence
Kalmijn AJ (1981) Biophysics of geomagnetic field detection. IEEE Trans Magnetics 17:1113–1124. Physical principles and behavioral evidence of motional-electric magnetic compass sense of sharks and rays and permanent-magnetic forced orientation in magnetic mud bacteria
Kalmijn AJ (1982) Electric and magnetic field detection in elasmobranch fishes. Science 218:916–918. Electrically evoked feeding attacks of *Mustelus canis* at prey-simulating dipole fields in shallow waters at 1-2 nV/cm and of *Prionace glauca* in the open-ocean waters. Orientation of *Urolophus halleri* in uniform electric fields at 5 nV/cm and in magnetic fields of earth strength
Kalmijn AJ (1984) Theory of electromagnetic orientation: a further analysis. In: Bolis L, Keynes RD, Maddrell SHP (eds) Comparative physiology of sensory systems. Cambridge University Press, Cambridge, pp 525–560. Comparison of linear and angular exploratory movements in support of an electromagnetic sense in animals
Kalmijn AJ (1988a) Electromagnetic orientation: a relativistic approach. In: O'Connor ME, Lovely RH (eds) Electromagnetic fields and neurobehavioral function. Alan R. Liss, New York, pp 23–45. Analogy between magnetic orientation of sharks and Einstein's example of relative motion between train and coil, by which he introduced his special theory of relativity
Kalmijn AJ (1988b) Detection of weak electric fields. In: Atema J, Fay RR, Popper AN, Tavolga WN (eds) Sensory biology of aquatic animals. Springer, New York, pp 151–186. Review plus approach algorithm and proof of ambient electric field orientation at sea
Kalmijn AJ (1995) Electric and near-field acoustic detection, a comparative approach. In: 80 jaar Vergelijkende Fysiologie, pp. 24–31, Tessel Offset, de Meern. Introduction of approach algorithm
Kalmijn AJ (1997) Electric and near-field acoustic detection, a comparative study. Acta Physiol Stockholm 161(suppl 638):25–38
Kalmijn AJ (2000) Detection and processing of electromagnetic and near-field acoustic signals in elasmobranch fishes. Phil Trans R Soc Lond B 355:1135–1141. Further analysis of electric and acoustic near fields, and introduction of ampullary equivalent circuit, amplifying signals by positive feedback driven by negative resistance of voltage-sensitive ion channels
Kalmijn AJ (2003) Sensing Nature's electric fields I: graded positive feedback in elasmobranch Ampullae of Lorenzini. In: Bezrukov SM (ed) Unsolved problems of noise and fluctuations,

vol 665. American Institute of Physics, pp 133–141. Ion channel noise in ampullary circuit? Intracellular recordings of electroreceptor cell in ampullae of Lorenzini

Kalmijn AJ, Blakemore RP (1978) The magnetic behavior of mud bacteria. In: Schmidt-Koenig K, Keeton WT (eds) Animal migration, navigation, and homing. Springer, Berlin, pp 354–355. Pulsed reversal and size of magnetic moment

Kalmijn AJ, Gonzalez IF, McClune MC (2002) The physical nature of life. J Physiol Paris 96:355–362. Validation of approach algorithm. Relevance of Einstein's special relativity. Ampullary circuit, featuring negative resistance driving positive feedback, and brain processing.

Keller BA, Putman NF, Grubbs RD, Portnoy DS, Murphy TP (2021) Map-like use of earth's magnetic field in sharks. Curr Biol 31:2881–2886

Kirschvink JL (1989) Magnetite biomineralization and geomagnetic sensitivity in higher animals. Bioelectromagnetics 10:239–259

Kirschvink JL, Walker MM, Diebel CE (2001) Magnetite-based magnetoreception. Curr Opin Neurobiol 11:462–477

Kirschvink JL, Winkelhofer M, Walker MM (2010) Biophysics of magnetite orientation: strengthening the interface between theory and experimental design. J R Soc Interface 7:S179–S191

Klimley AP (1993) Highly directional swimming by scalloped hammerhead sharks, *Sphyrna lewini*, and subsurface irradiance, temperature, bathymetry, and geomagnetic field. Mar Biol 117:1–22

Lissmann HW, Machin KE (1958) The mechanism of object location in *Gymnarchus niloticus* and similar fish. Exp Biol 35:451–486

Lohmann KJ (2010) Magnetic-field perception. Nature 464:1140–1142

Lohmann KJ, Johnsen S (2000) The neurobiology of magnetoreception in vertebrate animals. Trends Neurosci 23:153–159

Longuett-Higgins MS (1949) The electrical and magnetic effects of tidal streams. Monthly Notices of the Royal Astronomical Society, Geophysical Supplement, Vol. 5, No. 8. Mention of space charge distribution associated with motional electric field

Longuett-Higgins MS, Stern ME, Stommel H (1954) The electric field induced by ocean currents and waves, with applications to the method of towed electrodes. Papers in Physical Oceanography and Meteorology, Massachusetts Institute of Technology and Woods Hole Oceanographic Institution. Contribution No 690

Lorentz HA (1892) La Théorie électromagnetique de Maxwell et son application aux corps mouvants Leide, E. G. Brill. Original introduction to Lorentz force

Machin KE, Lissmann HW (1960) The mode of operation of the electric receptors in *Gymnarchus niloticus*. Exp Biol 37:801–811

Meyer CG, Holland KN, Papastamatiou YP (2005) Sharks can detect changes in the geomagnetic field. J R Interface 2:129–130

Molteno TCA, Kennedy WL (2009) Navigation by induction-based magnetoreception in Elasmobranch Fishes. J Biophys 2009:1–6

Montgomery JC, Bodznick D (1999) Signals and noise in the elasmobranch electrosensory system. J Exp Biol 202:1349–1355

Murray RW (1960a) The response of the ampullae of Lorenzini to mechanical stimulation. J Exp Biol 37:417–424

Murray RW (1960b) Electrical sensitivity of the ampullae of Lorenzini. Nature (London) 187:957

Murray RW (1962) The response of the ampullae of Lorenzini of elasmobranchs to electrical stimulation. J Exp Biol 39:119–238. Ampullae sensitive to electric currents along ampullary canals. Concurrent with Dijkgraaf and Kalmijn's 1962 behavioral proof of electrical sensitivity

Murray RW (1965) Elctroreceptor mechanisms: the relation of impulse frequency to stimulus strength and responses to pulsed stimuli in the ampullae of Lorenzini of elasmobranchs. J physiol (Lond) 180:592–606

Newton KC, Kajiura SM (2017) Magnetic field discrimination, learning, and memory in the yellow stingray (*Urobatis jamaicensis*). Anim Cogn 20:603–614

Newton KC, Kajiura SM (2020) The yellow stingray (*Urobatis jamaicensis*) can use magnetic field polarity to orient in space and solve a maze. Mar Biol 167:36

Newton KC, Gill AB, Kajiura SM (2019) Electroreception in marine fishes: chondrichthyans. J Fish Biol 95:135–154

Nimph S, Nordmann GC, Kagerbauer D, Malkemper EP, Landler L, Papadaki-Anastasopoulou A, Ushakova L, Wenniger-Weinzierl A, Novatchkova M, Vincent P, Lendl T, Colombini M, Mason MJ, Keays DA (2019) A putative mechanism for magnetoreception by electromagnetic induction in the pigeon inner ear. Curr Biol 29:1–8

Pals N, Valentin P, Verwey D (1982) Orientation reactions of the dogfish *Scyliorhinus canicula*, to local electric fields Netherlands. J Zool 32:495–512

Panofsky WKH, Phillips M (1962, 1972) Classical electricity and magnetism, 2nd edn. Addison-Wesley

Parker GH, van Heusen AP (1917) The responses of the catfish, *Amiurus nebulosus*, to metallic and non-metallic rods. Am J Physiol 44:405–420

Paulin MG (1995) Electroreception and the compass sense of sharks. J Theor Biol 174:325–339

Peters RC, Everts HP (1985) Frequency selectivity in the ampullary system of an elasmobranch fish (*Scyliorhinus canicula*). J Exp Biol 118:99–109

Peters RC, Eeuwes LBM, Bretschneider F (2007) On the electrodetection threshold of aquatic vertebrates with ampullary or mucous gland electroreceptor organs. Biol Rev 82:361–373

Petracchi D, Cercignani G (1998) A comment on the sensitivity of fish to low electric fields. Biophys J 75:2117–2120

Phillips (1996) Magnetic Navigation. J theor Biol 180(4):309–319

Rosenblum B, Jungerman RL, Longfellow L (1985) Limits to induction-based magneto-reception. In: Magnetite biomineralization and magnetoreception in organisms, pp 223–231

Serway RA (1990) Physics for scientists and engineers, 2nd edn. Saunders College Publishing, Philadelphia, PA

Tricas TC, New JG (1998) Sensitivity and response dynamics of elasmobranch electrosensory primary afferent neurons to near threshold fields. J Comp Physiol A 182:89–101

Tsong TY (1994) Exquisite sensitivity of electroreception in skates. Biophys J 67:1367–1369

Wald A (1947) Sequential analysis. Thorough Mathematical and Statistical Treatise, Dover

Walker MM, Diebel CE, Kirschvink JL (2003) Detection and use of the earth's magnetic field by aquatic vertebrates. In: Collin SP, Marshall J (eds) Sensory processing in aquatic environments. Springer, pp 53–74

Waltman B (1966) Electrical properties and fine structure of the ampullary canals of Lorenzini. Acta Physiol Scand 66 Suppl 264:1–60

GPSR Compliance

The European Union's (EU) General Product Safety Regulation (GPSR) is a set of rules that requires consumer products to be safe and our obligations to ensure this.

If you have any concerns about our products, you can contact us on ProductSafety@springernature.com

In case Publisher is established outside the EU, the EU authorized representative is:

Springer Nature Customer Service Center GmbH
Europaplatz 3
69115 Heidelberg, Germany

Batch number: 10732044

Printed by Printforce, the Netherlands